MENASHA CORPORATION

An Odyssey of Five Generations

Menasha Corporation

An Odyssey of Five Generations

Richard Blodgett

Greenwich Publishing Group, Inc.
Lyme, Connecticut

© 1999 Menasha Corporation. All rights reserved.

Printed and bound in the United States of America. No part of this publication may be reproduced or transmitted in any form or by any means, electronic or mechanical, including photocopying, recording or any information storage and retrieval system now known or to be invented, without permission in writing from Menasha Corporation, 1645 Bergstrom Road, Neenah, WI 54956, except by a reviewer who wishes to quote brief passages in connection with a review written for inclusion in a magazine, newspaper or broadcast.

Produced and published by Greenwich Publishing Group, Inc.
Lyme, Connecticut

Design by Clare Cunningham Graphic Design

Separation and film assembly by Silver Eagle Graphics, Inc.

Library of Congress Catalog Card Number: 99-60003

ISBN: 0-944641-35-0

First Printing: January 1999

10 9 8 7 6 5 4 3 2 1

TRADEMARKS:

The following are current or former trademarks of Menasha Corporation and its subsidiaries:

A Legacy of Strength through Vision™
Dripcut®
LEWISystems®
Orbis™
Promo Edge®
Sus-Rap®
Thermotech®
Tivar®
Traex®

PHOTOGRAPHY CREDITS:

page 82 appears courtesy of American Forest Products Industries

page 83 (lower right) appears courtesy of Coos Bay Times

page 37 appears courtesy of John Sensenbrenner

pages 19(left), 22(bottom), 26, 34(both), 41, 42(both), 44(lower left), 48, 58(top), 66, 75, 76, 80, 92, 95(upper right), 96, 98, 100 and 108 appear courtesy of the Menasha Public Library

pages 6, 43, 45, 49, 60, 62, 63, 65 and 102(both) appear courtesy of Neenah Historical Society

pages 18 and 33(both) appear courtesy of R. Mowry Mann

pages 21, 32, 39(both), 47(both) and 53(both) appear courtesy of Susan Gosin

All other photographs and historical items appear courtesy of Menasha Corporation and its subsidiaries. Photography work by Christopher Devlin Brown and Munroe Studios, Inc.

Image on preceding pages: oil by Richard Barrett depicting the arrival in Menasha of Elisha and Julia Smith

Table of Contents

Dedication	7
Chapter One "Most of Us Are Happy Driving Our Fords"	9
Chapter Two Elisha Smith, Rugged Pioneer	17
Chapter Three The Second Generation	37
Chapter Four Out of Woodenware, Into Corrugated	47
Chapter Five The Third Generation	63
Chapter Six Trees: A Family Heritage and Company Tradition	77
Chapter Seven Dick Johnson and the "Professionalization" of Menasha	91
Chapter Eight The Fourth Generation	107
Chapter Nine Plastics	123
Chapter Ten Completing the Package: Printing, Promotion and Corrugated Boxes	137
Chapter Eleven Menasha Corporation Today: Something to Be Proud Of	151
Family Tree	160
Timeline	162
Acknowledgments	165
Index	166

DEDICATION

This book celebrates the 150th anniversary of Menasha Corporation in 1999.

It is dedicated to Menasha Corporation employees, past and present, and the company's

Smith family owners, who have worked together for a century and a half to build

one of America's oldest and most successful privately held businesses.

The design element above each chapter title in this book is from

Menasha Wooden Ware Company's nineteenth-century letterhead on page 31.

AN ODYSSEY OF FIVE GENERATIONS

CHAPTER ONE

"Most of Us Are Happy Driving Our Fords"

This book tells the almost implausible story of the survival, growth and success of an American company through nearly 150 years of ownership by the same family, the descendants of Elisha D. Smith of Menasha, Wisconsin.

In an era when so many family-owned businesses — the Bingham newspaper chain, the Harry Winston jewelry store empire and the Dart Group (controlled by the Haft family), to cite just three examples — have been torn apart by disputes about money and management succession, the Smith family has marched to the beat of a different drummer.

Members of the Smith family do not always see eye to eye. (What family does?) Personalities sometimes clash. Yet, the Smiths have learned the fine art of how to disagree without destroying their company or their family. They share an old-fashioned Midwestern sensibility that values friendship, family ties and respect for others more than money. Disagreements, when they have occurred, have never been allowed to escalate into warfare.

Few family companies survive beyond the second or third generation of ownership. Moreover, it is rare that family branches stay close and remain focused on a common objective into the fourth generation. The Smiths have kept their company together for five generations, now heading into the sixth, with no end of family ownership in sight — a remarkable accomplishment.

A Legacy More Precious Than Jewels

Their company, Menasha Corporation, got its start when family patriarch Elisha D. Smith journeyed to the Wisconsin frontier in the mid-nineteenth century in search of opportunity, purchasing a bankrupt wooden-pail factory. That tiny factory evolved into today's Menasha, a profitable, growing, $1 billion-a-year manufacturer of paper, plastics, printed materials and corrugated containers, with significant holdings

At left, members of the Smith family gathered in Smith Park in 1997 to celebrate its 100th anniversary. The park was donated to the city of Menasha by their ancestor, Elisha D. Smith. Left to right are Mowry "Trip" Smith III; his father, Mowry Smith, Jr.; Mowry, Jr.'s cousin, Oliver Smith; and Oliver's son, Pierce Smith.

of prime timberland in the Pacific Northwest.

Menasha is owned by 130 of Elisha Smith's descendants, who have consistently rebuffed all proposals to sell their company. (In addition to the Smith family shareholders, stock is held by several non-family senior executives, retired executives and directors.) In 1977, when the Menasha board of directors considered a proposal that the company sell all or part of its stock in a public offering, Mowry Smith, Jr. (a great-grandson of founder Elisha Smith), told the board he had contacted most shareholders and that they considered their Menasha shares "an heirloom" and "they would rather part with the family jewels than with their stock." The board voted the proposal down.

The cohesiveness of the Smith family, together with Menasha Corporation's long-term perspective and ability to adapt to change, have given the company incredible staying power. As Menasha celebrates its sesquicentennial, it joins only a handful of companies that have managed to remain private — and to remain in family hands — for 150 years.

The Smith family owners are a fascinating group who have, almost to a person, inherited Elisha's ethic of independence and hard work. Because they view Menasha Corporation as a heritage to be preserved and passed on to their children, they have avoided the temptation to milk the company for all it's worth.

While being a Menasha shareholder

Nancy Des Marais, a great-granddaughter of Elisha Smith, and her granddaughter, Emily Vought, a sixth-generation Smith, pose next to the Elisha D. Smith plaque dedicated to the public library in Menasha.

provides a nice source of income, maximizing the company's earnings and dividends has never been the sole priority. Speaking about the generally relaxed attitude within the family about money, Katharine G. "Kig" Gansner, a fifth-generation descendant who is an attorney in Madison, Wisconsin, says with only a bit of overstatement, "Most of us are happy driving our Fords."

Professional Leadership

A common reason why family companies are sold is a lack of interest by the next generation of owners in managing the business. The Smith family has dealt with this challenge by hiring professionals to manage the company for them.

Through the fourth generation of Smiths (Elisha's great-grandchildren, now mostly retired), male descendants generally worked for the company. But those days are past. Of the nearly 50 fifth-generation descendants (Elisha's great-great-grandchildren), just one — William Shepard, group president, Menasha Packaging — works at Menasha Corporation. He heads the company's corrugated container operations, its largest business.

Other fifth-generation Smiths, now in their 30s, 40s and 50s, have struck out on their own, pursuing careers in law, business, publishing, education and the arts. For instance, Harry F. Jones III is an entrepreneur who owns and operates a chain of bookstores in upstate New York. Donald C. "Buzz" Shepard III is a partner at Faegre & Benson, a major law firm in Minneapolis.

"MOST OF US ARE HAPPY DRIVING OUR FORDS"

Menasha's directors were photographed at their December 1998 meeting. They are (clockwise from left): William Shepard, John Lauer, Evan "Van" Galbraith, Harry "Hank" Jones III, Katharine "Kig" Gansner, Clark Smith, Timothy Shepard, Thomas Prosser, Bruce Schnitzer, Curt Smith, Richard Clarke, Oliver Smith, Kirby Dyess, Donald "Buzz" Shepard and Robert Bero.

Curtis N. Smith is a director of private markets for a pension consulting firm in California. Well-educated and independent-minded, other members of the fifth generation enjoy equally successful careers.

The company began "professionalizing" its management — that is, hiring experienced executives from outside the family — in 1961 when Richard L. Johnson, recruited from Marathon Corporation, became Menasha's president and chief executive officer. Johnson headed the company for 20 years and is regarded as one of its legendary leaders. Menasha's president and CEO today is Robert D. Bero, who joined the company in 1979 as vice president in charge of the Plastics Division and was elected CEO in 1993. Of the company's seven operating groups, six are headed by non-family executives, the one exception being packaging, Bill Shepard's domain.

Family members and professional management work in tandem guiding the company. Descendants of Elisha Smith are welcome and encouraged to join the company. However, career advancement is based on merit, not genealogy, unlike years ago when the presidency was generally handed down from father to son.

As management has shifted from family members to professionals, the board of directors has changed also. The Menasha board once consisted almost entirely of family members, but there was a growing recognition beginning in the 1970s that the company needed outside perspective to survive and prosper in a highly competitive business environment. Today the board is comprised one-half of Smiths and one-half of non-family executives. The outside directors include individuals with broad business experience, such as Kirby A. Dyess, vice president and general manager of new business development for Intel Corporation, and Bruce W. Schnitzer, former president and chief executive officer of Marsh & McLennan, Inc.

Working To Uphold a Tradition of Integrity and Concern For People

Despite being removed from day-to-day operations, the fifth generation of Smiths cares deeply about the company. Like earlier generations, they take great pride in Menasha Corporation and feel strongly that it should be a responsible employer and active corporate citizen.

A priority of the Smith family is to make today's sixth-generation youngsters feel connected to Menasha Corporation. This children's history book of the company, published in 1994, is part of that effort.

Elisha Smith was known for his honesty, integrity and concern for people, values that continue to guide the company. "It may sound corny, but we really do believe in fairness and ethical conduct," says Donald C. "Tad" Shepard, Jr., a fourth-generation Smith who was Menasha's president and CEO in the 1980s.

In 1995, at the insistence of family members, Menasha increased its charitable contributions to 1.5 percent of pretax income from one percent. Moreover, the fifth generation is not at all shy when it comes to prodding the company on corporate issues of personal concern. At the urging of Gansner, Menasha made a conscious effort in the mid-1980s to recruit the first women for its board of directors. Today, Gansner is herself a Menasha director, one of two women on the board.

One of the challenges now faced by the Smith family and Menasha Corporation is to get the sixth generation (Elisha's great-great-great-grandchildren) to appreciate why Menasha Corporation is so special.

There is great hope within the Smith family that at least some sixth-generation youngsters will one day consider careers at Menasha, and that they will in any case feel, as their parents and grandparents feel, that Menasha Corporation is a family legacy to be sustained and nurtured. Timothy C. Shepard, a fifth-generation Smith, puts the issue succinctly. "Unless some of our children work for the company," he says, "they won't care. And if they don't care, Menasha will not survive as an independent company."

Bringing the Sixth Generation Into the Fold

There are about 70 sixth-generation Smiths, mostly in their teenage years or younger. Unlike their parents and grandparents, who grew up learning about the business, the sixth generation is growing up in locations across the United States with little connection to Menasha Corporation. In most cases, their parents don't work for Menasha, so the youngsters do not hear stories about the family business around the dinner table. Considerable effort is being expended to whet the sixth

"MOST OF US ARE HAPPY DRIVING OUR FORDS"

The company sends birthday presents that familiarize young Smiths with its products. Nick Gansner, left, a sixth-generation family member who graduated from Dartmouth in 1997, displayed this model truck from Menasha. Peter Shepard, another sixth-generation Smith, wrote the letter below to thank the company for a present.

generation's interest in Menasha Corporation and the Smith family heritage.

Since the early 1990s, the company has sent birthday presents each year to all the sixth-generation children with the objective of familiarizing them with the business that will one day be their charge. The presents are Menasha products, such as plastic shipping boxes that a youngster might find nifty because of the way they fold into different shapes.

In 1994, Menasha published a hardcover, beautifully illustrated children's history book of the company, distributed to all youngsters in the Smith family.

The annual Smith family picnic, begun in 1991, is another effort to make the sixth generation feel connected not only to the company but also to the family. The picnic was the idea of Curt Smith, the California pension consultant, and is organized and managed by three of his fifth-generation cousins, Anne Des Marais, Kig Gansner and Mowry "Trip" Smith III, who live in Wisconsin.

Held each spring at Smith Park (donated to the city of Menasha a century ago by Elisha Smith), the picnic attracts approximately 100 of Elisha Smith's descendants

PETER SHEPARD

Dear Mr. Bero,
Thank you very much for the boxes. I never knew that Manasha made that kind of box. I put all of my stuff in them so know my room and my closet are really clean. I really like them. Well maybe I will join Manasha Coppiration some day. Well thanks again-for the boxes.

From,
Peter

AN ODYSSEY OF FIVE GENERATIONS

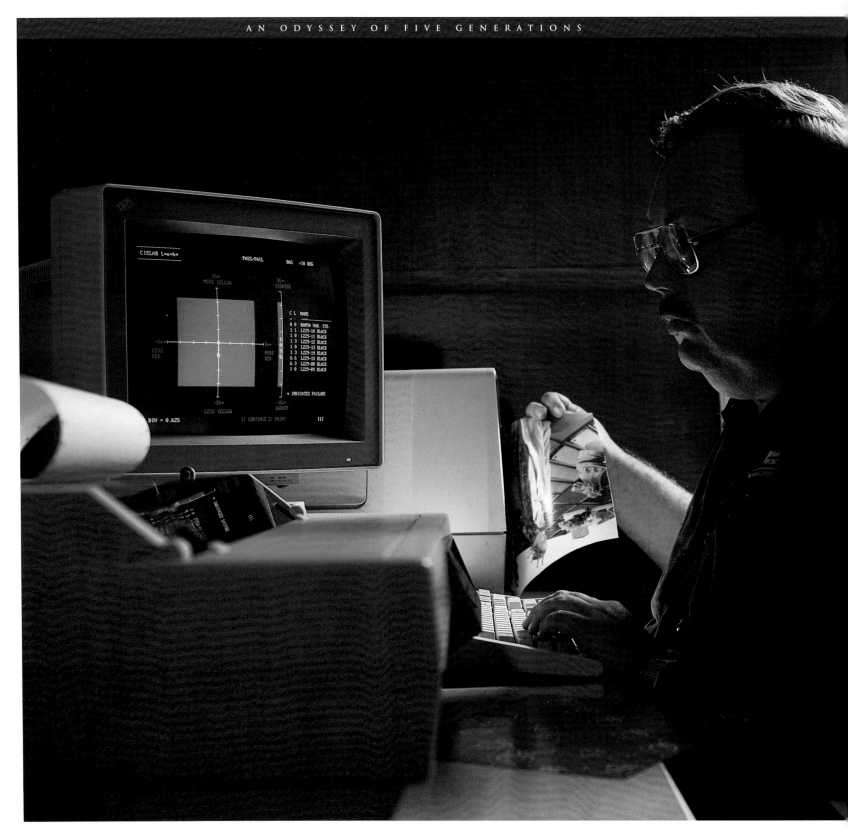

Menasha is today a growing multinational corporation with skilled people employing modern technology. In the Neenah Printing Division, an advanced specrophotometer, left, helps measure printed colors against a preset standard to ensure the quality of the packaging and promotional material it prints for customers. A state-of-the-art registration system for mounting printing plates, above, guarantees precision and consistency.

who journey to Wisconsin from their homes across the United States. The older Smiths have a grand time renewing acquaintances. But the best part, according to Trip Smith, is that the youngsters get together, playing soccer or just hanging out, and in this way meet their extended family of cousins with whom they will one day share ownership of the company. "The importance of early bonding cannot be overstated," Trip Smith says. "By knowing each other when they are young, they will find it easier to understand each other and get along when they grow up. And they will be less likely to want to sell the company."

Setting the Company's Financial Sights Higher

At the same time, Menasha's management recognizes that continued family ownership is desirable, but will only be possible if the company generates reasonable financial returns.

Through the 1950s, the same family members who ran the company were also its directors and its largest shareholders. The payoff from being a member of the Smith family was to work for the company and earn a salary. Dividend payments were small, so those Smiths who did not work at Menasha did not have much of a financial stake. They enjoyed owning a business, but their investment was not lucrative financially.

That changed dramatically in the 1980s, when Menasha grew rapidly and began to pay significant dividends. Suddenly, all the shareholders, whether they worked for the company or not, took notice and began to expect a reasonable return. This is a transition that many family companies go through, some successfully, some not. In the case of Menasha, the transition has been successful, in part because the shareholders are tolerant in those years when financial results are weak. They understand that Menasha participates in cyclical businesses, such as paperboard and timber, and results are bound to vary from year to year.

CEO Bob Bero's goal is to provide returns comparable to those of the Standard & Poor's 500 Index. To that end, Menasha has embarked upon an aggressive growth strategy to capitalize on opportunities in such markets as corrugated containers, returnable packaging, specialty plastics and in-store promotional graphics and to improve the performance of each of its businesses. This strategy recognizes that markets and customer needs change and Menasha must adapt with new products and production processes. Although the strategy has taken time to generate positive results, like the transitions many companies go through when they implement new strategies, it now appears to be taking hold and results are on an upswing.

With uninterrupted ownership for nearly 150 years by the same family, Menasha Corporation is surely one of the most unusual companies in America. Let's take a closer look at Menasha: how it started, how it grew, the challenges it faces today and what its story tells us about entrepreneurship in America.

CHAPTER TWO

ELISHA SMITH, RUGGED PIONEER

In the fall of 1850, when Millard Fillmore was president, 23-year-old Elisha Dickinson Smith and his 21-year-old bride, Julia Ann Mowry Smith, packed their belongings and headed west for the Wisconsin frontier.

Elisha was a native of Brattleboro, Vermont, a commercial center on the Connecticut River — one of seven children in a family that traced its ancestry on his mother's side to the *Mayflower*. As a young man, he was energetic and restless. On completing high school at age 17, he traveled to Boston to work as a clerk in a wholesale distribution firm and then moved to Woonsocket, a textile-manufacturing city in Rhode Island, where he found a job in a dry-goods store. It was in Woonsocket that he met Julia, the daughter of a prosperous local banker.

When Elisha and Julia fell in love and became engaged, Elisha could have joined his future father-in-law's bank and been financially secure the rest of his life. But the idea of being a banker and working for his

*E*lisha Smith was 40 when he posed for this photograph in 1867. The circa 1900 view, opposite, shows part of Menasha Wooden Ware's stave yard in the foreground and downtown Menasha in the background. The light-colored building on the right is the Elisha D. Smith Library, a gift from Elisha to the the city.

father-in-law did not appeal to him in the least. Prior to his marriage, he visited Atlanta, Georgia, deciding he and Julia would live there and he would open a dry-goods business to be his own boss. However, on returning to Woonsocket, he found a letter awaiting him. This letter would change Elisha and Julia's lives forever. Written by an acquaintance, Dr. Doane, who had just moved to the remote village of Menasha, Wisconsin, the letter urged Elisha to visit Menasha before deciding on his future. Intrigued, Elisha traveled to Menasha in July 1850, liked what he saw and returned to Woonsocket to marry Julia. Their wedding was held on Thursday, October 24, 1850. They left for Wisconsin the very next morning.

Elisha's father-in-law, Spencer Mowry, the Rhode Island banker, never forgave Elisha for taking Julia nearly half a continent away. Mowry provided financial backing for Elisha's ventures in Menasha, presumably to assure that Julia would be comfortable financially. Even as he did so, however,

Spencer Mowry, Elisha Smith's father-in-law, was a tough-minded, self-made businessman and prosperous Rhode Island banker. Reared on a farm, and with little formal education, Spencer Mowry began a construction business in Woonsocket, Rhode Island and became one of the city's wealthiest and most influential residents. In 1844, at age 41, he was elected president of the city's Globe Bank, continuing in the position until his death. He was also Woonsocket justice of the peace for more than 40 years and was elected three times to the state general assembly.

he claimed continually for the next 37 years, until his death in 1887, that Elisha was a failure. He even insisted that Elisha was "the least qualified to carry on business of any man that I ever saw that has done as much business as he has." This was written when Elisha was in his early 50s and nearing his greatest success as an entrepreneur!

Elisha did experience his share of setbacks — most people who start a business do. But he ended up creating a large, successful enterprise and is today considered one of Wisconsin's important early industrialists. Elisha's older son, Charles, eventually wrote to Mowry (his grandfather) asking that he stop subjecting Elisha to "insult after insult." Mowry wrote back that Charles should mind his own business. Mowry's endless hostility must have been trying for Elisha, who addressed letters to his father-in-law "Dear Father" — and got back letters addressed "Mr. E. D. Smith" or "Dear Sir."

The Arduous Journey to Menasha, Wisconsin

It is easy to understand why Elisha, a strong-minded individualist who wanted to be in business for himself, would be willing to try his luck in a remote settlement in a brand-new state (and, in the process, get away from his rancorous father-in-law). But it is remarkable that Julia, raised in luxury, would be willing to move to the Wisconsin frontier. Of course, women of that time were expected to follow their husbands, and she headed off to Menasha without complaint and perhaps with a sense of excitement and adventure.

Their 1,000-mile journey from Woonsocket to Menasha took more than a week and was filled with twists and turns. It was a trip not many of us today would find agreeable, although by mid-nineteenth-century standards they traveled in style.

Railroads were still in their infancy, and there was not yet any direct rail service between New England and the Midwest. First, the newlyweds traveled by train (on five separate lines, no less, each requiring a change of cars and transfer of baggage) to Buffalo, where they boarded a Lake Erie steamship which took them to Detroit. From Detroit, they traveled by train across the state of Michigan to New Buffalo, where they boarded a steamer for Chicago, then a town of 18,000 without paved streets. They continued by steamer up Lake Michigan to Sheboygan, Wisconsin, where they embarked for Fond du Lac by horse-drawn carriage — a 40-mile trek which took two days because the dirt roads were in such poor condition. From Fond du Lac, they traveled north on Lake Winnebago by boat to Oshkosh and then changed to a sternwheeler, the *Peytona*, which took them to Menasha. However, their trip ended on a sour note. When the *Peytona* tried to land at Menasha, the town's irascible founder, Curtis Reed, refused permission because he was angry with the captain. While the *Peytona* lay at anchor offshore, a smaller craft came out to fetch Elisha and Julia. On reaching land, there was "not a person in sight, and we

Elisha and Julia Smith traveled to the small frontier community of Menasha in 1850 aboard a sternwheeler. The vessels were still a feature of Menasha's waterfront 65 years later, above. The painting, right, depicts the Pail Factory in 1856. The plant is along the right bank of the canal, and the village is on the left.

made our way as best we could through the mud to our hotel," Elisha later recalled.

The newlyweds were part of a great mid-nineteenth-century migration into Wisconsin. Many of the newcomers were German immigrants who had left their native land because of crop failures and the collapse of democratic movements in Prussia and Austria. Between 1840 and 1850, Wisconsin's population soared an amazing 10-fold, then rose another 150 percent in the ensuing decade, reaching 775,000, according to the 1860 census.

Menasha was one of many tiny settlements springing up like wildflowers across the state. Situated on the western shore of Lake Winnebago by the Fox River, Menasha was founded in 1848 and was the outgrowth of a community first known as Winnebago Falls. As it leaves the lake, the river has two branches which merge to the west. Between the branches lies Doty Island, once home to a village of Winnebago Indians. The Winnebagos had ceded the island to the federal government in the early 1800s, moving west to present-day Portage, Wisconsin. When Elisha and Julia arrived, Doty Island was sparsely populated and covered with dense woods. There was a single dirt road across the island connecting small residential communities immediately to the north and immediately to the south — today's twin cities of Menasha and Neenah.

In 1850, the year of their arrival, the federal government completed the construction of a lock and canal on the northern branch of the river, a defining moment in the city of Menasha's history. These facilities allowed boats to travel the length of the Fox River from Lake Winnebago to Green Bay and, from there, to Lake Michigan. The building of the lock and canal, together with the region's boundless forests, spurred Menasha's rapid early growth in such industries as paper, timber and woodenware. It was perhaps the building of the lock and canal, and the likelihood that an economic boom would follow, that attracted Elisha to Menasha.

Elisha and Julia lived initially in a small two-room cabin with a smoke-pipe through the roof for a chimney, and he became a partner in the dry-goods store of his acquaintance, Dr. Doane. An early advertisement listed an inventory of fabrics, clothing, groceries, hardware, crockery and other household items.

As in any new community on the outer edge of civilization, Menasha had its share of rascals and con artists, a point Elisha was soon to learn the hard way. To his dismay, the dry-goods business languished from the start, and Doane departed for points unknown, leaving Elisha with all the partnership's debts. A Dun & Bradstreet agent, hired to track Doane down, concluded that he was "one of those migrating, speculating sort of gentlemen" who never stayed long in one place. Doane was never found, nor was he heard from again.

But that was not the worst of it. Within a matter of months, Elisha had been hoodwinked repeatedly by customers to the tune of several thousand dollars. One episode in the winter of 1850-1851 involved a steamboat, the *Berlin*, being built locally. The owner persuaded Elisha to sell him merchandise on credit, promising to pay in the spring. When spring arrived and the owner defaulted, Elisha secured a lien on the boat and became its owner. He then arranged for a man named Malbourn to sell the boat in La Crosse. Malbourn sold the vessel, but kept the money. Elisha then

Julia Ann Mowry Smith, Elisha's wife, was bright and outgoing. She gave up a life of luxury in Rhode Island to go with him to Menasha, and she never regretted having done so. She and Elisha remained devoted to each other throughout their lives.

hired a lawyer to sue Malbourn. The lawyer won the judgment, collected $2,000 — and pocketed the entire amount himself. The lawyer died not long thereafter, and Elisha never did receive a penny. "Thus ended my first steamboat experience, and not a very lucrative one for me at that," Elisha later recalled wryly.

Shortly thereafter, Elisha lost $2,500 when the builders of another boat failed to pay for goods despite repeated promises to do so. And he lost $1,000 when he extended credit to the builders of a wood-surfaced road from Menasha to Appleton, just to the north. (That highway, long since paved, is known today as Old Plank Road.)

Years later, he said, "My early experience here in storekeeping was exceedingly trying, when in my ignorance I supposed everybody was honest." Because Elisha was fastidious about keeping promises and paying debts, he found it difficult to comprehend others who did not. Nonetheless, he could be generous even to those who disappointed him. Once, coming upon a man who owed him $1.87, which was overdue, Elisha demanded payment right then and there. Elisha seldom took no for an answer, and the man had little choice except to comply. However, seeing the man was in financial straits, after collecting the $1.87 — and satisfied the debt was now repaid — Elisha gave the man $5.00 and told him to use it for his family. Back then, $5 was enough to meet the monthly rent with something left over for groceries.

Even as he struggled in his business

In the 1890s, as their older son Charles began to assume responsibility for managing the company, Elisha and Julia found time for travel, including this trip to Egypt. They are riding the two camels at far left.

because of his generous and honest nature, those same qualities made him a leader in the community. Elisha had a natural ability to inspire the trust and confidence of others. When the people of Menasha decided to build a Congregational church, they called on him to become a church trustee and help lead the fund-raising drive despite the fact that he was only 24 years old and not a churchgoer himself. In 1857, Elisha and another man recruited a minister, the Reverend H. A. Miner, for the church. As recalled later by Miner, Elisha told him, "We are not either of us members of the church but it seems to us a shame for a town like this with a population of 2,000 to be without a resident pastor, and we have come to induce you to accept the pastorate." Impressed by Elisha's "frankness and sincerity of expression," Miner accepted. He and Elisha became lifelong friends, and years later Miner's son, Willis, served as president of Menasha Wooden Ware Company from 1921 to 1936.

Elisha also became active in local politics, initially as a Whig and later as an anti-slavery Republican. There was great contention in Menasha during the 1850s between the Whigs (the national party of Daniel Webster and Henry Clay, which favored western expansion, among other causes) and the Democrats (the national party of James Buchanan and Stephen A. Douglas, the "establishment" party that dominated the federal government). Nationally, both parties were deeply divided over the burning issues of slavery and states' rights that increasingly dominated national debate.

Elisha, not averse to fisticuffs now and then, became so embroiled in the discord between the two local parties that he slapped the Democratic owner of the *Menasha Advocate* in outrage over some newspaper article. The next morning, Elisha was arrested for assault and immediately put on trial. As it so happened, at the hour of his arrest he and his fellow Whigs were about to embark on an outing on Lake Winnebago and were anxious to get started, trial or no trial. Moreover, Elisha's friends did not want to leave without him. So they simply pushed their way en masse into the courtroom and shoved him out the door. Testimony was suspended, and the trial was never resumed. No one seemed to mind in the least that justice had been frustrated in this manner.

Menasha Corporation's Origins: The Pail Factory

Although Elisha Smith is considered the founder of Menasha Corporation, in fact the company had been started even before he and Julia arrived in Wisconsin.

In 1849, when Elisha and Julia were still young lovebirds in Woonsocket, Rhode Island, three men — Nathan Beckworth, Joseph Sanford and C.W. Billings — opened a small factory in Menasha to manufacture wooden pails. Their entire investment was less than $1,000, part of it borrowed at 50 percent annual interest. (No wonder they failed!) The Pail Factory, as it was called, had a single lathe, and the three men did all the work themselves. Menasha Corporation's beginnings date to this rudimentary mill.

In and of itself, the idea of going into the woodenware business was sound. Throughout the nineteenth century, barrels and other wooden packaging were the dominant means of shipping all kinds of goods — candy, wheat, crackers, sugar, pickles, nails, etc. — to market. Woodenware was a large and growing industry. As the population increased and

The drawing above depicts the delivery of barrels to a customer in the company's earliest years. By late in the nineteenth century, as the business prospered, the company used large delivery wagons like the one below. The driver is an employee named Joe Beaudo.

rail transportation improved, demand for wooden packaging expanded dramatically.

Nonetheless, the three men could not make a go of their new business, even though they had the benefit of vast local supplies of timber and water power from the Fox River. After a year, having produced only 1,500 pails and having failed to earn a profit, the partners sold the factory to Joseph Keyes, Norman Wolcott and Lot Rice. The new owners "met with similar success," according to *The History of Winnebago County* by Richard Harney.

It was Elisha who took this tiny, struggling operation and built it into something meaningful. Looking to operate a second business in addition to his dry-goods store, he bought the factory in 1852 for $1,200 with his father-in-law's financial support. Typical of many entrepreneurs who start small, he initially ran the plant as a one-man shop, operating the machinery himself and delivering the products by horse-drawn wagon.

With Elisha at the helm, the business began to grow. Within four years, he had hired his first employees and was selling products in markets as far away as Chicago. His primary product was, by now, barrels for the shipment of wheat. Then came the Panic of 1857, which drove thousands of American businesses into bankruptcy. This would be a turning point for Elisha. He not only survived the panic, but withstood pressure from his father-in-law to return to Rhode Island. Visiting from Woonsocket in 1858, Spencer Mowry found Elisha's business to be doing poorly. He urged the Smiths to move back to Woonsocket and offered Elisha a job at the bank. As Elisha recalled these events, Spencer Mowry told him, "You can't succeed here. You will wear yourself out and have nothing to show for it. Better sell out for what you can get and take the cashiership of our bank." According to Elisha, "I replied that I would not go back if he *gave* me the bank."

Early on, Elisha recognized that one of the keys to the success of his woodenware business was to get a railroad to come to Menasha. By the mid-1850s, railway tracks were crisscrossing the nation at a frantic pace. Communities that were connected to a railroad could import raw materials and ship out finished products with ease —

Elisha and Julia built this large Victorian home at the corner of Keyes and Park Streets on Doty Island. Most other local industrialists lived outside the city, but Elisha and Julia stayed because they loved the city and its people.

much faster and less expensively than by river. Communities that were bypassed by the railroads faced the prospect of economic strangulation because they would be outside the loop of national commerce.

Together with other leaders of the community, Elisha worked long and hard in lobbying for a railroad to be built from Menasha to Manitowoc, a port on Lake Michigan, but this idea collapsed with the onset of the Panic of 1857. Then in 1861, the Chicago & Northwestern Railroad — known familiarly as the Northwestern — announced plans to extend its line up the western side of Lake Winnebago from Fond du Lac to Appleton, bypassing Menasha and Neenah because it did not want to take on the expense of building bridges across the twin cities' network of river branches and canals. Recognizing the urgency of the situation, a local delegation traveled to Chicago to meet with representatives of the Northwestern. At that meeting, a deal was struck: the Northwestern agreed to run a branch line to Menasha and Neenah, in return for which the communities agreed to pay for the bridges, railroad ties and grading of the line. The Northwestern's branch was completed in 1862, just as Elisha's business embarked on a period of dramatic expansion spurred by the Civil War. It was also in 1862 that Elisha closed his dry-goods store to concentrate on the manufacture of woodenware.

The Pail Factory became a leading supplier of storage and shipping containers to the Union forces during the war. By war's end, the plant — now the largest woodenware factory in Wisconsin — had more than 100 employees and was producing 900 barrels, 300 tubs, 2,400 butter tubs, 300 washboards and 300 boxes of clothespins a day, a far cry from the ramshackle little operation Elisha had acquired a decade earlier.

Elisha's personal life was also humming along quite nicely, as he and Julia built a large Victorian home in the fashionable part of town and began to raise a family. They had five children: Mary, born in 1853; Charles in 1855; Henry in 1857; Carrie in 1867; and Jane in 1872. Only three of the children lived to adulthood. Mary died in infancy. Carrie died at age 12, a particularly harsh blow for Elisha and Julia.

ELISHA SMITH, RUGGED PIONEER

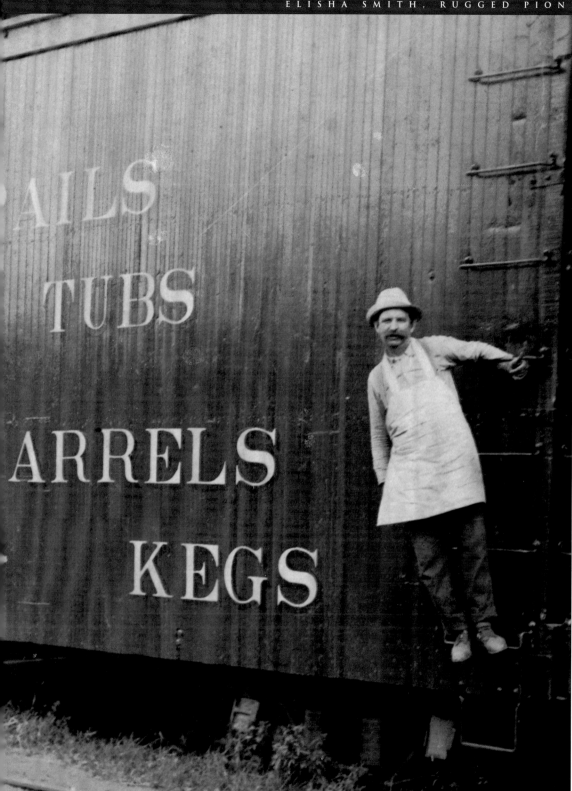

Beginning around the turn of the century, Menasha shipped its products in oversized boxcars, 10 feet high versus the normal 8 feet, purchased from a defunct circus. The cars were painted green, with yellow and red lettering, and became a familiar sight on rail lines across the nation. The boxcar shop repair crew included Foreman William Mechler, far left, and Henry Wilpolt to his immediate left.

Fighting Fires

Because Elisha was manufacturing wooden products in buildings constructed of wood, fires were a constant peril. Once a fire started, it might quickly engulf all the wood in its path — factory and contents alike. The first fire occurred in March 1856, destroying one of the buildings where lumber was dried. The main plant went up in smoke six years later. Fire struck again in 1870, 1878 and 1890, each time destroying all or part of the manufacturing complex. In a sense, these fires were a blessing in disguise: in each case, a bigger and more modern factory rose from the ashes. Following the 1890 fire, Elisha opted for brick construction, although, as we shall see in later chapters, conflagrations kept occurring anyhow.

Finding Timber

Elisha also began to search for new sources of raw material. The Pail Factory initially obtained wood to make its products from the forests of Menasha and surrounding communities. In those days, loggers seldom replanted what they cut. Once the trees were gone, the loggers moved on to a new forest and left the old land to regenerate itself. America's resources were believed to be limitless, and in a sense, they were, given the country's relatively small population and its vast stands of timber.

As the supply of local wood was depleted, Elisha cast an ever-wider net. By the 1860s, he was purchasing timber from the Wolf River basin about 25 miles west of Menasha. To bring this timber to the factory, huge rafts of logs were towed down the Wolf River and across a series of interconnecting lakes: Lake Poygan, Lake Butte des Morts and Lake Winnebago. Even into the early years of the twentieth century, the towing of log rafts to the factory was a source of excitement in the community (not unlike a circus coming to town) and attracted many onlookers, particularly children.

Elisha eventually looked still further afield, purchasing a large tract in Portage County, nearly 80 miles west of Menasha. By 1890, he had acquired land and timber rights throughout Wisconsin, Michigan and Minnesota. Moreover, he hired lumbermen known as "cruisers" to extend the search. Charles Worden was one of the legendary cruisers in the company's employ. Born in upstate New York, he joined the company in the 1870s and became one of the great pedestrians of all time, walking thousands of miles of forest in Wisconsin, Michigan, Minnesota, Oregon and Idaho looking for timberland the company might acquire. "Walking," Worden once said, "is absolutely the best exercise. It costs nothing. Just get a good pair of walking boots and your expense is at an end. The best cure-all for every kind of disease and ailment is steady walking." Worden lived to 86, working — and walking — right up to the day he died in 1930.

Worden was one of many exceptional employees who spent their entire careers with the company. In fact, it was common

Menasha owned vast stands of Sitka spruce in the Pacific Northwest, using the wood to make butter tubs. Butter could be stored in a Menasha tub for up to 12 months without molding, compared with only six months in competitors' tubs made of ash.

for employees to stay with the company 40 to 50 years or even longer.

In the Service of Christ

Even though Julia was a devout Christian all her life, Elisha had relatively little interest in religion in his younger years. This changed suddenly in 1858, when he had a spiritual awakening at age 31. While in Chicago on business, he visited a Baptist church and was overcome with evangelical fervor.

Elisha was never one to do anything halfway, and that was certainly true of religion. Returning to Menasha, he threw himself into Christian devotion and good deeds with joyful passion — and continued to do so the rest of his life. He became so involved with the First Congregational Church of Menasha that some people started referring to him as "the man who built the church and ran it." Unlike many other local manufacturers, he closed his factories on Sundays and even sent horse-drawn wagons through town to gather up anyone needing a lift to church, whether they were heading for the First Congregational Church or any other.

Moreover, he became a paragon of Christian concern for those in need. The *Neenah Citizen and Menasha Register*, in a 1998 special edition celebrating the history of the twin cities, pointed out that Menasha and Neenah became a center of millionaire industrialists in the second half of the nineteenth century, many earning their fortunes in the region's booming paper industry. Yet, of the initial wave of local industrialists, including at least eight millionaires, "only Elisha D. Smith gave large sums of money to public causes," according to the newspaper. Elisha was also one of the few local industrialists to reside in the city. For those reasons, Elisha was the most popular and influential of the early industrialists, even though his wealth was dwarfed by the fortunes of some families in the paper business.

Elisha led the fund-raising drive to build the First Congregational Church of Menasha even though he was not a church-goer himself at the time. He simply wanted to do his part to improve the city. Later, following a religious awakening, he became so active in the church that he was called "the man who built the church and ran it."

Beginning in 1858, Elisha maintained a locked ledger itemizing every contribution he made to a needy individual or worthy cause. Over the next 41 years, that ledger grew to 250 pages, containing thousands of entries — from major gifts to colleges and religious institutions to countless small outlays such as "a poor Frenchman who had lost his leg, $6.00," and "caring for and burying a sick stranger, $15.00." As his giving to needy individuals and worthy causes increased, in 1884 Elisha hired an agent to investigate charities in an effort to ensure the most effective use of his philanthropic dollars.

Elisha sometimes came up with ideas for good deeds in the most unconventional ways. Traveling in Germany, he observed the women of that country toiling in the

Elisha believed in the value of education and provided financial support to dozens of colleges and universities. Wisconsin's Ripon College, which he served as a trustee, named this building Elisha D. Smith Hall in his honor. It continues to bear his name today.

fields. It was the middle of the summer, and his thoughts turned immediately to Menasha, some 5,000 miles away. Before retiring that evening, he wrote to his secretary, directing that a cruise ship be chartered at his expense to provide the mothers of Menasha with a one-day outing on Lake Winnebago. Visiting a town in California, he was struck by the beauty of its shade trees — and promptly ordered 50,000 saplings for distribution to the people of Menasha.

His support of missionary work was exemplary. He became a member of the executive committee of the Wisconsin Home Missionary and served as a director of the American Board of Commissioners for Foreign Missions. For more than a decade, he was volunteer superintendent of a mission school southwest of Menasha. He traveled to the school each Sunday after morning services at the First Congregational Church. Years later, a former student wrote, "He was intensely in earnest about the salvation of the members of the school. He often took one and another aside for a little quiet talk and urged them to give themselves to the service of Jesus. I remember so well his talk with me, and I am sure that he was the means in the Lord's hands of my conversion when I was only a little girl."

It was not long before Elisha became a close friend of Dwight L. Moody, the famous nineteenth-century evangelist. Smith and Moody first met in 1859 when the latter visited Neenah to hold a Sunday school convention. Two years later, Moody abandoned a successful business career at age 24 to devote himself full time to missionary work in Chicago. Moody was an inspirational preacher who spoke in glorious terms of God's love and mercy, just the type of upbeat message that appealed to Elisha. As their friendship blossomed, Elisha contributed heavily to Moody's educational and missionary programs.

Elisha was no mere Sunday Christian. Religion became an integral part of his daily life. He sometimes ended meetings with suppliers or customers by launching into a discussion of the joys of Christian service to the community. In later years, he distributed leaflets containing practical wisdom about Christian living. He even placed these leaflets in the bottoms of pails and tubs shipped from his factory in hopes they would be read by customers.

One of Elisha's interests was to help finance the building of churches. Over his lifetime, he gave money toward the construction of nearly 40 houses of worship, mostly in Wisconsin, belonging to various denominations. Education, especially Christian education, was another passion. He gave a total of $22,000 (a princely sum in those days) to various colleges — from Ripon, Lawrence and others in Wisconsin to such far-flung institutions as Grinnell and Wilton in Iowa, Tuskegee University in Alabama and several Christian universities in Turkey. Elisha D. Smith Hall at Ripon College was named in his honor, as was the Elisha D. Smith Gymnasium at Beloit College.

A Lifelong Love Affair

In addition to being unique among the early business leaders of Menasha in his large donations to worthy causes, Elisha was unusual in his open affection for his wife, something uncommon in that period. His love and faithfulness never faded, nor did hers.

On their 25th anniversary, Elisha and Julia wore their wedding clothes to a reception held in their honor at the National Hotel. On their 30th anniversary, he publicly declared that he loved Julia even more than he did the day they were married.

Julia was herself smart and down-to-earth. The couple made a great team. While walking across a bridge in 1885, Elisha and Julia were attacked by two bandits, one of whom hit the 58-year-old Elisha in the head with a sandbag, knocking him down. As the two men jumped on Elisha and tried to rob him, Julia punched one of them in the eye with the end of her parasol and shouted for Elisha to get out his revolver. The robbers backed off a few paces, then resumed their assault when it became apparent that Elisha did not have his revolver with him. Meanwhile, Elisha had picked up a board that had broken from the bridge in the fracas, and he with the board and Julia with her parasol succeeded in driving the robbers off.

How Elisha Nearly Lost the Company

There came a time when Julia's love and emotional support helped carry Elisha through a humiliating experience, one that nearly cost him his company. In 1872, the Pail Factory employed 250 and was the largest woodenware manufacturer in the Midwest, perhaps in the United States. But expenses were rising faster than income, a fact unknown to the public, since financial results were kept strictly private.

On April 6, 1872, the *Menasha Press* announced some astonishing news: the Pail Factory, the community's largest employer, was $250,000 in debt. Unable to meet payments, it had closed its doors, throwing its employees out of work. The factory reopened not long afterwards in receivership.

Public attention focused immediately on Elisha. As often happens in these cases, rumors arose that he had done something imprudent or fraudulent to bankrupt the business. "There were many stories circulated that reflected upon his integrity, which stung him to the quick," Miner, the church pastor, said later. Elisha eventually repaid all the Pail Factory's debts in full, even those which had been discharged in receivership, and his good name was restored.

Spencer Mowry ultimately came to his son-in-law's rescue with an infusion of cash. The business was incorporated under the name Menasha Wooden Ware Company, with Spencer Mowry owning 90 percent of the stock. Julia owned two percent, Elisha none. Henry Hewitt, Sr., president of a local bank, was elected president of Menasha Wooden Ware, while Elisha — continuing to run the business as general superintendent — became a salaried employee at $1,200 a year. It would be nine years before Elisha regained the presidency.

Perhaps wiser and more experienced as a result of his company's brush with disaster, Elisha emerged as a more capable manager. Never again did he let the company get into financial trouble.

Growth and Innovation in the 1880s and 1890s

As the 1880s began, the entire region of northeast Wisconsin was advancing rapidly into the industrial age. The nation's first hydroelectric plant was commissioned in Appleton in 1882. Three years later, Menasha Wooden Ware became one of the first corporations in the United States to install its own electric generator, beginning the company's conversion from water power to electricity.

A cache of letters, discovered recently in the attic of one of Spencer Mowry's descendants, provides a fascinating picture of Menasha Wooden Ware in the period from 1878 to 1887. These letters were written by Spencer Mowry, Elisha Smith, Elisha's son Charles and others associated with the company.

One of the major themes of these letters is the importance of the Fox River and how much flow it produced. In a letter written in 1879, Charles worried that the water level of the river was receding, thereby reducing the rate of flow and limiting the amount of power available to the factory. "You know that all streams, as the timber is cleaned up on their headwaters and tributaries, gradually get lower and lower," he wrote. Menasha began to employ steam-generated electricity six years later.

Elisha's letters talk repeatedly about strong demand for the company's

Barrel production continued well into the twentieth century. These barrels are moving along a conveyor in the factory for loading into a boxcar.

Above, workers pose outside the factory in 1895. At right is the ornate letterhead from Menasha's stationery in the nineteenth and early twentieth centuries.

woodenware, but the inadequacy (in his view) of prices. They also offer a rare glimpse at the efforts of an industry trade association to limit competition and bolster prices, practices that were common and perfectly legal at the time. In January 1885, Elisha wrote that three new woodenware factories had "very foolishly" opened for business the prior year in Michigan, even as the Woodenware Manufacturers Association paid four others a total of $50,000 to close their doors for a year. "It is a big load to carry for the Association," he wrote, "but it is very much better than to have the prices go to pot."

The letters are also filled with squabbles between Elisha and his father-in-law over money, land and other issues. Elisha borrowed routinely from banks in Menasha and Chicago to finance inventories and purchase timberland. In 1879, Spencer Mowry warned Elisha, "If you keep on the way you have been going, the debts will be so large it will swamp you." This time, however, unlike his earlier brush with bankruptcy, Elisha managed debt carefully. Bank borrowings helped build the company and enabled it to buy land that years later proved to be extremely valuable.

The 1880s and 1890s were a golden era for Menasha Wooden Ware Company. As its sales increased, the company bought more land and built new facilities. In 1886, it purchased 85 acres on Doty Island to provide additional drying yards for lumber and staves. In 1888, it opened a three-story paint shop, where products were painted by a crew of 80. The company's manufacturing complex also included, among other facilities, a saw mill, machine shop and cooper shop, all located on the northern channel of the Fox River. In addition, the company owned stave-making and saw mills at Auburndale, Stevens Point, Wrightstown and Apple Creek, Wisconsin, mostly acquired or built in the 1880s. Local timber was cut into staves at these mills, and the staves were then shipped to Menasha to be assembled into products. Staves were less expensive to ship than timber.

The company also began to step up its marketing effort, opening a sales office in Chicago. In 1886, William Boyd, a Chicago native, was hired to head the office, remaining with the company 44 years until his death in 1930.

Candy pails became one of Menasha's major product lines. In fact, the company eventually held a dominant share of the candy pail market because of a patented lock that prevented pilferage. These pails were used by candy manufacturers to ship licorice, anise squares, gum drops, butterscotch buttons, cinnamon balls and the like to retail stores, and by the stores themselves to hold and display the candy. Because other woodenware manufacturers' candy pails lacked this lock, they were a natural target for anybody with a sweet tooth. How tempting it must have been for shipping clerks, railroad personnel and others to reach in and grab a handful when nobody was looking. A shipment of candy in one of

Menasha became the world's largest producer of candy pails, a market it dominated because of a patented lock that prevented pilferage. It was said that William Boyd, who headed the company's Chicago sales office, sold more candy pails than any other person in the history of the United States.

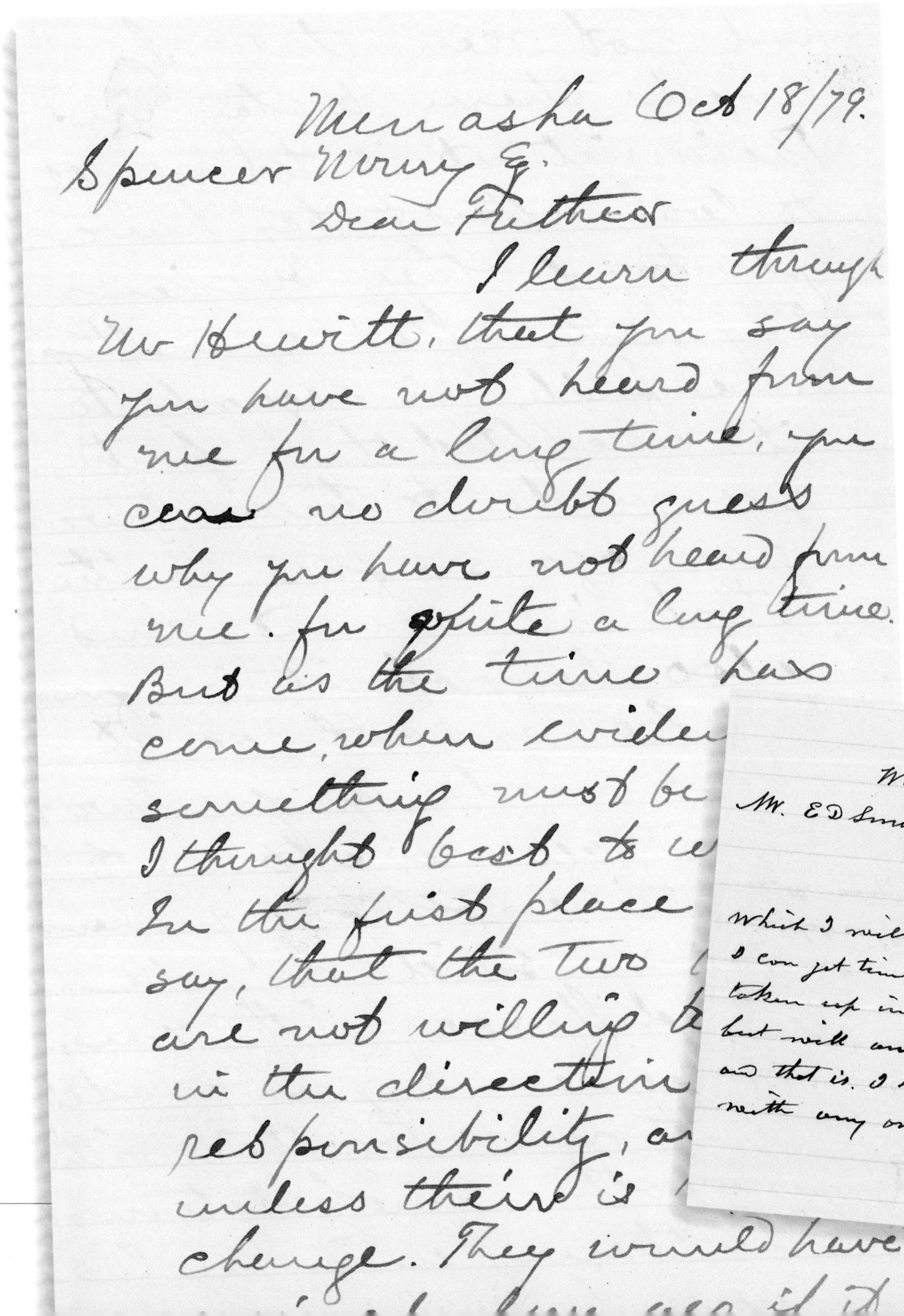

Menasha Oct 18/79.

Spencer Mowry Eq.

Dear Father

I learn through Mr Hewitt, that you say you have not heard from me for a long time, you can no doubt guess why you have not heard from me, for quite a long time. But as the time has come, when evide[ntly] something must be [done] I thought best to w[rite] In the first place [I] say, that the Two [Companies] are not willing t[o go on] in the directio[n of] responsibility, a[nd] unless their is [a] change. They would have...

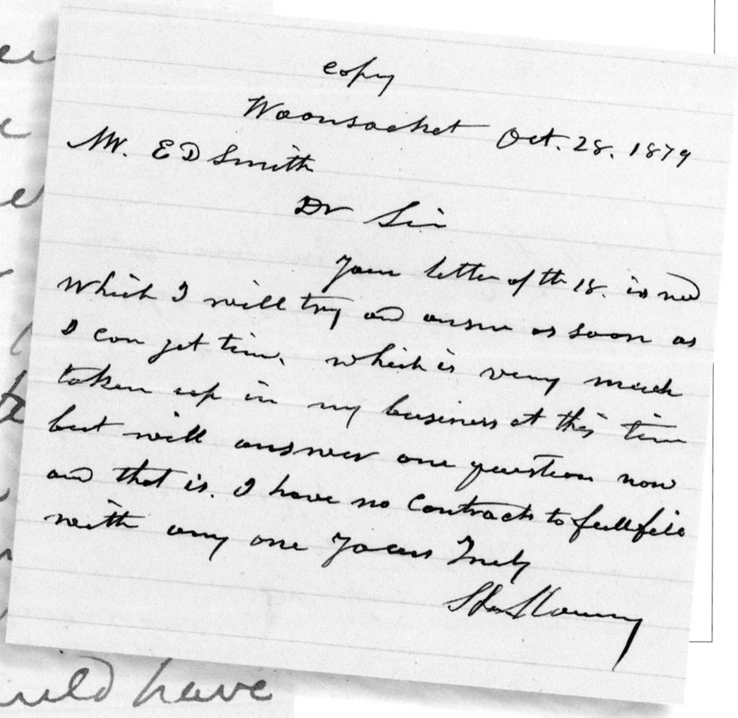

copy

Woonsocket Oct. 28. 1879

Mr. E D Smith

Dr Sir

Your letter of th 18. is rec[eived] which I will try and answer as soon as I can get time, which is very much taken up in my business at this time but will answer one question now and that is. I have no Contracts to fullfill with any one Yours Truly

Sp Mowry

Elisha and his father-in-law, Spencer Mowry, who lived in Woonsocket, Rhode Island, kept in touch by letter. In 1879, when these letters were written, Spencer Mowry was the principal owner of Menasha Wooden Ware, having acquired a 90 percent interest when he rescued it from bankruptcy seven years earlier. The Mr. Hewitt referred to in Elisha's letter is Henry Hewitt, Sr., president of the National Bank of Menasha, whom Spencer had installed as Menasha's president to watch over Elisha. Spencer Mowry not only kept the letters he received from Elisha, but also hand-wrote and kept copies of his own replies — hence, the word "copy" on the letter below.

Elisha's many gifts to the city of Menasha included the Elisha D. Smith Free Public Library, left, and Smith Park, opposite and below. Some 6,000 residents, joined by four marching bands, paraded to his home to thank him for the park. He told the crowd, "I think any person will live longer and happier in this world by doing something to help mankind."

these competing products might arrive at its destination with half the candy gone! Given Menasha's superior product and commanding market position, it was said that William Boyd, the Chicago office manager, sold more candy pails than any other person in the history of the United States.

Meanwhile, day-to-day management of Menasha Wooden Ware Company was passing into the hands of Elisha's older son, Charles. Indeed, by the late 1880s, Elisha was devoting less and less time to the company and more to travel and his philanthropic activities, becoming the greatest benefactor the city of Menasha has ever known.

Elisha spent 10 years assembling a 25-acre site on Doty Island to try to attract Downer College to the city. When his offer was rejected, he gave the land to the city of Menasha in 1897 for a park — the same park where the Smith family now holds it annual picnic. Until he made his gift, there were no public parks in Menasha and at most 30 acres of parkland in Neenah, according to the *Neenah Citizen and Menasha Register*. A newspaper of the period described the community's response to the gift as follows:

"After the unanimous acceptance of the proposition by the council, an adjournment was had, and then began one of the liveliest and heartiest impromptu celebrations ever witnessed in Menasha. The fire and church bells rang, the whistles blew, sky rockets were sent up, fire crackers added to the din, tin horns were tooted, gongs were pounded, and everybody made merry over the grand gift. It was nearly midnight before the noise wholly subsided."

Elisha was astonished, having feared the

The gazebo in the park was renovated in 1997 with financial support from the Menasha Corporation Foundation.

city might reject his gift because of conditions he had attached to it, including a provision that the drinking of alcoholic beverages not be allowed in the park, a prohibition that continues to this day. To his even greater surprise, the celebration resumed the following Saturday night, when some 6,000 residents of Menasha and Neenah, joined by four marching bands, paraded to the home of the 70-year-old Elisha to thank him again for the park. Addressing the crowd, Elisha is said to have declared, "Let me say to you, my friends, that it is a grand thing to do for those who are needy. I think any person will live longer and happier in this world by doing something to help mankind."

Elisha also donated $32,500 to purchase a site and erect the building for what became known as the Elisha D. Smith Free Public Library. It was opened in 1898 and continued until 1971, when it was replaced by a new facility.

End of an Era

By the 1890s, Menasha Wooden Ware Company was the largest employer in the city of Menasha and Elisha had become the grand old gentleman of the community. He was not only the city's greatest benefactor, but also one of the few original settlers of Menasha still alive. When he died in 1899 at age 72, following a brief battle with cancer, residents of the city treated his passing as if he had been an exalted potentate. Flags were flown at half-mast, buildings were draped in black and businesses closed their doors for the funeral. The funeral procession extended more than a mile.

Julia died two years later in 1901.

Beginning half a century earlier, and surviving many setbacks, Elisha D. Smith had built his company into the world's largest manufacturer of turned wooden products, including barrels, pails and tubs. At his death, the name Menasha Wooden Ware was known nationwide. Revenues were $1 million annually, a good-sized business for the era, with 1,000 employees and an annual payroll in excess of $350,000. Every day, more than 150 Menasha Wooden Ware boxcars traveled the rails, delivering the company's products to customers across the country.

It would be up to Elisha's older son, Charles, to take the company to even greater heights in the woodenware business.

AN ODYSSEY OF FIVE GENERATIONS

CHAPTER THREE

The Second Generation

Charles Robinson Smith is remembered today as one of the most astute chief executive officers in Menasha Corporation's history and one of the most prominent Wisconsin businessmen of his era. He was a kind of one-man band who dominated the company's management for two decades, leading it to sustained growth and prosperity during the last great boom period of the woodenware business.

Industrialists like Charles Smith helped forge the modern American economy, propelling the nation to global economic leadership in the late nineteenth and early twentieth centuries. "Obviously, for Elisha Smith to come all the way out here to Menasha in the 1850s was a gutsy move," says Mowry Smith, Jr., a great-grandson of Elisha and grandson of Charles. "Somebody had to get the business started. But I think Charles R. Smith was really the dominant builder of this company."

Charles, or C. R., as he was known, led an interesting and productive life. He loved running the family company. A contemporary described him as being "always at work." Yet, Charles was anything but a one-dimensional industrialist who could not see beyond his job. He was very much a family man. Moreover, he had a wonderfully pungent wit. In February 1879, he wrote to his grandfather, the formidable Spencer Mowry, "I judge if you are carrying on lawsuits that you are well this winter." Not many people dared speak to Spencer Mowry in this manner. But Charles did, and got away with it, because of his natural forthrightness and lack of fear of anyone or anything.

Charles believed in the dignity of a good day's work and detested idleness. He could be tough and relentless. It was his custom to stop at a local barbershop each morning for a shave before going to the office. One time, an employee happened into the barber shop, not realizing Charles was there, and complained there was no work at the pail factory that morning because the staves were wet. Charles is said to have sprung from the barber's chair and proclaimed, "No work? Then what in thunder are you planning to do today?"

Charles R. Smith, Elisha's older son, was Menasha's president from 1899 to 1916. Opposite is turn-of-the-century pail-making at the company's plant.

Charles's son-in-law, Donald Shepard, Sr., reclines on the lawn of his home with his wife, Sylvia Smith Shepard, and friends. Donald Shepard was a Menasha vice president known for the imaginative advertising he wrote for the company. The Shepards lived on Forest Avenue, next to Sylvia's uncle, Henry Smith.

Education and Early Years In the Business

Charles was born in Menasha in 1855 and attended public school in the city. In an era when most business executives were not college educated, he went to Princeton University, earning a bachelor's degree — a testament to Elisha's belief in the importance of a first-rate education. Charles made many friends at Princeton, including several who went on to become the heads of large companies. Charles maintained contact with them all his life and even bought a house in New York City, in addition to his home in Wisconsin, so he could visit regularly with his many East Coast friends. Menasha was the leading supplier of wooden packaging to business, and high-level contacts of this sort were a terrific sales tool.

Another of his college friends was Woodrow Wilson, who entered Princeton as a freshman when Charles was a senior. Charles and the future U.S. president remained cordial even after college, often seeking each other out at Princeton reunions.

Graduating in 1876, Charles returned to Wisconsin to join the family business. A born entrepreneur with boundless energy, he not only worked in the family company but also started his own business, a broom-handle factory on the Menasha Wooden Ware Company premises. Menasha Wooden Ware was just coming out of receivership when Charles returned home, and perhaps he wanted to hedge

Taken aback, the employee replied that he would go to a local mill and get some firewood to take home. Charles is said to have rejoined, "Save your shoe leather. I'll see that a load is delivered to you."

When the man returned home, he found a cord of oak logs which took him two days to split. One of Charles's favorite sayings was, "He who cuts his own wood warms himself twice."

THE SECOND GENERATION

Charles Smith was the epitome of the successful turn-of-the-century American businessman. His Queen Anne-style home, built in 1891, still stands on Forest Avenue.

his bets by having a business to fall back on in case Menasha faltered. Because Menasha did not then manufacture broom handles, there was no conflict between his personal business and the family company. Indeed, Elisha helped finance Charles's factory, indicating the former was supportive of the latter's desire to be his own boss at least part of the day.

Splitting time between the two operations, Charles was elected secretary of Menasha Wooden Ware in 1878, just two years after having graduated from college, and was named to its board of directors two years after that. Charles's personal business was successful in its own right and was eventually acquired by Menasha in 1894, with Charles receiving 900 shares of Menasha common stock valued at $225,000. Charles was, by now, de facto chief executive officer of Menasha Wooden Ware Company even though his father remained titular head as president.

Elisha's Three Children: Charles, Henry and Jane

Charles and his younger brother, Henry, had entirely different personalities. Elisha was proud of them both. In 1885, when Charles was 30 and Henry 28, Elisha wrote to a friend, "I am very fortunate in having two sons so capable and competent to succeed me in the business." Elisha's comments notwithstanding, it was Charles who possessed the leadership skills and drive to manage the company.

Henry started out on virtually the same educational and career path as his older brother. He graduated from Princeton in 1878, two years after Charles. On receiving his bachelor's degree, he returned to Wisconsin to work in the family business, as had Charles. In 1881, he became Menasha Wooden Ware's corporate secretary, succeeding Charles, who became treasurer.

In family businesses where there are two capable sons, it is not uncommon that they end up at loggerheads competing to run the company. Charles and Henry, who lived next door to each other, were rivals to some degree. "Our local historian dug out some material which showed that if Uncle Charles added a new porch, then Henry had to have one," says Nancy Des Marais, Henry's granddaughter. "Or if Henry built a new room, Uncle Charles had to do the same, only bigger and better."

However, when it came to the family business, there was no dispute. Henry deferred to Charles, avoiding a battle. Henry retired in his mid-40s and thereafter filled his life with world travel, continuing to collect a salary without having to work for it. This arrangement apparently satisfied both brothers — and averted any quarrel that might have destroyed the company.

Des Marais says she adored Henry, her grandfather, because he was "so different and interesting and charming." She describes him as a handsome, debonair man who rode about town in a horse-drawn coach and, in later years, always drove a fancy automobile. She adds, "He took trips to Asia and came home with gorgeous gifts from China, India, Burma. That's how he liked to spend his life. He wasn't much of a homebody."

Henry also became something of a black sheep in the family when he was the first Smith to get a divorce. "People in that era were very strict," Des Marais observes. "Divorce was not accepted." Des Marais says Henry's children seldom talked to him and rarely mentioned his name after he left their mother to marry another woman.

However, the divorce apparently did not affect Henry's relationship with Charles. As far as we know, the two brothers remained on friendly terms all their lives.

The third sibling was their younger sister, Jane. She married a man also named Smith, and married again after her first husband died. In the custom of the time, she devoted herself to civic activities and did not take part in the company's management even though she inherited 12 percent of its stock.

Charles owned 45 percent of the shares, acquired through inheritance and the sale of his business to Menasha, and also controlled the 21 percent still in their father's estate. This majority ownership cemented his position as boss. Henry's ownership was 19 percent. The remaining shares were held by three employees: F. D. Lake, Menasha's treasurer; John D. Schmerein, corporate secretary; and Willis Miner, a senior executive who became president in the 1920s. Each owned about one percent.

A Man of Action

Taking charge formally as president in 1899 (he had, in fact, already been running the company for about a decade), Charles grew the business rapidly. Aggressive expansion was his policy.

In 1900, seeking new sources of wood, he purchased timberland and built a stave mill at Warner in northwestern Wisconsin. Warner had originally been called Flambeau Falls, but the residents had changed the name to Warner in tribute to an executive of the Wisconsin Central Railway when it extended its line to the town. When Charles built the stave mill, they changed the name again — this time to Ladysmith, in honor of his new wife, Isabel. The town continues to be known as Ladysmith today.

In 1905, Charles acquired timberlands in Crandon, Wisconsin, about 100 miles north of Menasha. A rail line did not exist to transport timber from that part of the state to Menasha, and the Chicago & Northwestern Railroad turned down his request to build one. Never one to let adversity stand in his way, Charles founded a railroad, the Wisconsin Northern, nicknamed the Whiskey Northern, to lay the track. Unfortunately, the railroad encountered all kinds of operating problems and

THE SECOND GENERATION

Pails were stacked inside Menasha's boxcars according to a particular pattern to maximize the number that could be loaded into a car.

consistently lost money, turning out to be one of his rare blunders. It was sold after his death.

Charles repeatedly expanded the company's production facilities at Menasha in response to growth in demand for woodenware. In 1901, he doubled the size of the paint shop, and in 1904, he added a fifth story to the pail factory. As the company grew, its manufacturing complex eventually encompassed 50 buildings and more than 100 drying houses on a 65-acre site.

Charles was relentless in seeking opportunities to bolster sales and profits. When a traveling circus went out of business, he seized on the chance to acquire its oversized boxcars, converting them to ship Menasha products. Railroads at that time charged a base rate to ship a boxcar full of goods regardless of weight. This was costly for Menasha. Its shipments, consisting of empty barrels and tubs, were unusually light because, in the words of one Menasha executive, they contained "a lot of air." The oversized boxcars helped alleviate the problem: more product could be shipped in each boxcar without having to pay additional freight charges.

Capitalizing further on these extra-spacious vehicles, Charles turned them into giant mobile billboards. They were repainted bright green, and these words were emblazoned on their sides in large yellow-and-red letters, arranged in the following manner:

MENASHA	PAILS
WOODEN WARE CO.	TUBS
ESTABLISHED 1849	BARRELS
MENASHA WIS	KEGS

Menasha's colorful boxcars became a familiar sight gliding along the rails all across America, bringing the company and its products to the attention of the public.

Investment in Western Timberlands

Perhaps Charles's greatest and most lasting contribution was to extend the company's timber holdings to the Pacific Northwest.

Although some wondered why a Midwestern company would go so far afield for raw materials, Charles recognized the need to diversify Menasha's sources of supply — and also realized that western timberlands were then selling at incredibly low prices.

He began in 1903 by acquiring land and timber rights in northern Idaho, a rough-and-tumble backwater. Just four years earlier, federal troops had put down a political uprising by miners at a site not far from Menasha's newly acquired property. Then in 1905, the Idaho governor who had called out the troops was assassinated. Despite this climate of violence and political turmoil, the timber industry was beginning to invest in the state and provide much-needed jobs. Charles was one

Right, Otto Beaton (closest to the camera) posed proudly in June 1926 when he became foreman of the pail factory. He had joined Menasha 16 years earlier as a lathe operator. Below is a barrel factory crew.

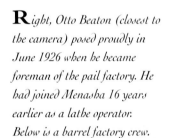

For 18 years after graduating from Princeton in 1876, Charles Smith not only worked for Menasha Wooden Ware but also ran his own business making brooms and wooden handles. The experience of managing a successful, growing company of his own was enormously valuable when Charles became president of Menasha.

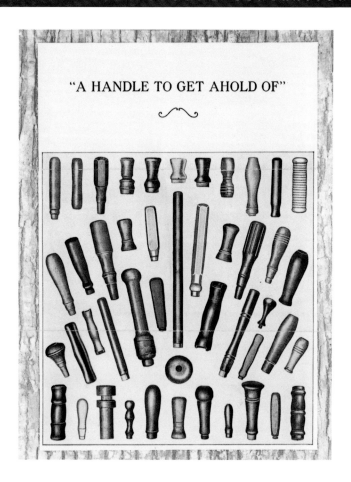

"A HANDLE TO GET AHOLD OF"

of the first to see opportunity in Idaho, buying timberland at a time when many others still shied away.

His acquisitions in Idaho were followed by purchases in Oregon, Washington and Saskatchewan, Canada. In 1904, he acquired an interest in a sawmill at Empire, Oregon, by purchasing ownership in the Southern Oregon Company, a major timberland owner in the region. Two years later, he opened a western office at Wallace, Idaho. All told, some 20,000 acres of the prime West Coast timberland that he acquired early in the century are still owned by the company.

Speaking of his grandfather Charles, Mowry Smith, Jr., says, "My goodness, he went out there and bought timber at ungodly low prices. It took real vision." Menasha's land holdings in the Pacific Northwest are today one of the jewels in the corporate crown, providing an ongoing source of revenue from the sale of timber.

Charles Outside the Business

In appearance, Charles was the epitome of the successful turn-of-the-century businessman — dignified-looking, conservatively dressed in a suit and high starched collar, always neatly groomed.

He continued many of the family interests and traditions his father had begun. Like Elisha, Charles was a parishioner of the First Congregational Church and gave generously to charity. He had a lifelong interest in education, serving for a time as the city of Menasha's volunteer superintendent of schools and sitting for many years on the boards of Beloit and Lawrence colleges.

In 1889, when he was 34, Charles married Jennie Mathewson of Menasha. In 1891, they had a son, Mowry, and one year later another son, Carlton. In 1895, just six years after their marriage, Jennie died at age 32 giving birth to their daughter, Sylvia. As often happens in times of grief, the extended Smith family pulled together. Henry's wife, Ella, stepped forward to help care for the children. "My grandmother, bless her heart, was sort of a second mother to the children until Charles remarried, and even after that," Nancy Des Marais says. Five years after Jennie's death, Charles married Isabel Rogers of Neenah.

Charles's sons both attended Princeton, his alma mater, and joined the company on graduation, eventually managing the business. His daughter, Sylvia, graduated from Smith College in Northampton, Massachusetts. She returned to Menasha, got married, had two sons and became a leader in the community. "My mother would have been a big part of the company," says her older son, Tad Shepard. "She was a real pistol. She was like her dad. She had his business ability and smarts, but women just didn't work in business at that time."

Elisha's Descendants: the Charles Line and the Henry Line

Charles's strong leadership influences the company even to this day.

Charles and Henry both had children. Their sister, Jane, did not. In many family-owned businesses, one family branch or another predominates — and that was true of Menasha Corporation after Charles's death. Because Charles owned a majority of the stock, his children inherited ownership control as well as the reins of management. His son, Mowry Smith, Sr., was CEO from the mid-1930s through the early 1960s, and his grandson, Tad Shepard, in the 1980s.

Henry's descendants, by contrast, did not generally go into the business, although they continued to own stock (and still do) and have been represented on the Menasha board of directors. The branches have always gotten along fine.

Though for many years the Charles branch ran the company, today the family and the company recognize that neither branch has a corner on management talent. Consequently, the family is making a concerted effort to get sixth-generation youngsters in both the Charles *and* Henry lines interested in the company in hopes they will consider careers at Menasha Corporation.

Storm Clouds on the Horizon

Charles died unexpectedly at the height of his career. In March 1916, while riding on a train near Spokane, Washington, visiting Menasha's timberlands and West Coast sawmills, he suffered a seizure. He was taken to New York for medical treatment but never recovered. He died on May 12, 1916, at age 61.

Henry outlived Charles by 25 years, dying in 1941 at 84. Their sister, Jane, passed away that same year at age 69.

Charles's 17-year presidency would prove to be Menasha's last great hurrah in the woodenware business. By the time of his death, the woodenware market was already entering upon a precipitous decline that would test the leadership and ingenuity of the next generation of Smiths and the staying power of the company itself.

THE SECOND GENERATION

Left, butter tubs are loaded into boxcars. Far left, the job of moving staves from the yard to the factory was tough physical work. Below is a cooper's hammer used for more than 40 years by Menasha employee Elmer Massey. Every cooper made his own hammer, sloping and balancing it to his individual requirements. This hammer was presented in 1970 to the Menasha Historical Society by Clem Massey, Elmer's son and is now kept at Menasha's public library.

CHAPTER FOUR

OUT OF WOODENWARE, INTO CORRUGATED

Companies are like people — they go through both good and bad times. In fact, virtually every company in the world has endured at least one difficult period in its history. Successful companies are able to survive the bad times, rebound and change.

And so it has been with Menasha. Riding high in the woodenware business when Charles Smith died, Menasha was suddenly swept into one of the most wrenching periods in its history. The tough times would last more than two decades — from 1916 right through the Roaring Twenties and on into the Great Depression of the 1930s.

Many companies would have been bankrupted by the prolonged difficulties experienced by Menasha. However, Menasha survived through the sheer doggedness of its Smith family owners and its many loyal employees who refused to give up. "There's no doubt the company had its ups and downs," says Donald

Charles's two sons were Carlton Smith, above left, and Mowry Smith, Sr., above right. Both joined the company after graduating from Princeton. Opposite is the bottom shop in 1915. It made the bottoms for Menasha pails, barrels and tubs.

Turner, Jr., whose father, Donald Turner, Sr., joined Menasha after World War I and spent his entire career as a member of management. "But they had a lot of good employees. And if you have good employees and good supervisors and you are led in a reasonable direction, that goes a long way for a company's survival."

As it turned out, Menasha did more than survive. During this pivotal phase in its history, the company transformed itself from a business rooted in products of the nineteenth century to a new company with entirely new products and with hope for the future. Without the changes that took place in the 1920s and 1930s, Menasha would not exist today.

Filling the Management Gap

Menasha's problems began when the death of Charles Smith ushered in a period of management turmoil. Charles was a talented businessman who had dominated the company's affairs for nearly three decades but

Carlton Smith, Mowry Smith, Sr., and Willis Miner, left to right, visited the company's timberlands in Oregon.

whose major failing was that he neither shared decision-making nor groomed a successor. Perhaps he had planned to remain in charge until his two sons were ready. However, the sons, Mowry, Sr., and Carlton, were still in their senior year at Princeton when Charles died, and neither was prepared to run the business.

Meeting in July 1916, the board of directors elected a longtime employee, Frank Lake, to succeed Charles as president and chief executive officer. Charles's brother, Henry Smith, became chairman of the board, remaining uninvolved in day-to-day management, as had been true for years. Willis Miner, another longtime employee, was elected vice president. Lake, Miner and Henry Smith were each paid $12,000 a year.

At that same meeting, Charles's sons, just two months out of college (and now, together with their sister, Sylvia, the majority owners of the company), assumed their first management positions. The older son, Mowry Smith, Sr., was elected to the board seat that had been held by Charles, while the younger, Carlton Smith, became assistant secretary. Their salaries were set at a more modest $1,200 a year, indicating their junior standing within the management team.

However, this junior management status did not stop Mowry, Sr., from pushing immediately for change. A likable extrovert who always spoke his mind, Mowry felt passionately about the family business and had strong opinions about how it should be managed. He thought the time had come for Menasha to diversify, and he proposed that the company begin producing corrugated boxes in addition to its line of woodenware packaging. Lake, the new president, disagreed, feeling there was too much competition in corrugated. That opposition cost Lake his job. He later recalled, "I told Mowry that if they went ahead, I'd resign, and he accepted it like that. You could have knocked me over with a stone." Menasha did not begin producing corrugated boxes until nearly a decade later, but Lake was nonetheless out in the wink of an eye. His presidency had lasted just 26 months. We do not know how well Lake and Mowry Smith, Sr., got along personally, but it seems obvious that Mowry did not

OUT OF WOODENWARE, INTO CORRUGATED

Willis Miner was Menasha's chief executive officer from 1921 to 1936, leading it through one of the most difficult periods in its history. Though not a member of the Smith family, he was very close to the Smiths and enjoyed their total confidence.

admire Lake's talents as company president.

Thomas M. Kearney, a 62-year-old attorney, succeeded Lake as president. A management consultant and partner in a Racine, Wisconsin, law firm, Kearney was well known for his ability to reorganize troubled companies. He came to Menasha in September 1918 and stayed 28 months, improving its manufacturing operations and reducing costs. Having accomplished those tasks, he resigned and returned to his law practice at the end of 1920.

Kearney was, in turn, succeeded by Willis Miner, the Menasha vice president mentioned earlier. Miner had a distinctive personality. He did not drink and was very frugal. One colleague described him as being "straight-laced." He dressed immaculately in nicely tailored dark suits. A University of Wisconsin graduate, he was smart and well educated. On top of all that, he was physically very strong and, on request, could tear a Chicago phone book in two.

Miner possessed the leadership skills Menasha needed so desperately in this period of uncertainty and change. Equally important, he enjoyed the absolute confidence of the Smith family, whose members knew him to be someone of honesty and integrity. Miner had grown up in Menasha, where his father, the Reverend H. A. Miner, was pastor of the First Congregational Church. The Smith and Miner families were very close. Willis Miner once said that "Elisha D. Smith was almost like a father to me" and that "Charles R. Smith treated me at all times like a younger brother."

In assuming Menasha's presidency, Miner restored long-term management stability, heading the company for the next 15 years.

Collapse of the Woodenware Market

These events occurred against the backdrop of a sharp and sudden decline in the market for woodenware packaging.

Packaging is such an integral part of everyday life, like electricity or automobile travel, that people often take it for granted without realizing how central it is to the economy and how it has changed over time. Throughout the nineteenth century, nearly all food products, and even items such as candles and nails, were shipped to market in barrels, kegs, tubs and other wooden containers. And, of course, Menasha was the world's largest producer of these bulk containers, which protected the products during shipment and were used to display them in stores.

However, as the twentieth century began, a packaging revolution was sweeping across America — a revolution that was bad news for Menasha. This revolution involved the sudden and dramatic shift away from bulk wooden packaging to new types of consumer-sized packaging made of paperboard, glass and other materials. In large part, this change reflected America's transition from a rural to an urban society, with greater emphasis on the needs of city residents. The new forms of packaging, though in some cases more expensive than bulk wooden containers, were more convenient for consumers and more sanitary. In addition, they made it easier for consumer product companies to establish brand identities. (As an example, Life Savers, introduced in 1913, has an identity not only because of its product shape and composition, but also because of its distinctive packaging.)

The opening shot in the packaging revolution was fired in 1896, when the National Biscuit Company introduced Uneeda biscuits in a box with a waxed paper liner, the first national brand packaged in that manner. The product was an immediate smash hit, and other companies soon followed.

The barrel business was already in decline when Charles Smith died in 1916. It was not until after his death, however, that the most devastating events occurred. Menasha's largest product line, by far, was candy pails, a market it dominated. With America's entry into World War I, that market disappeared virtually overnight. The war created demand for a new product, the individually wrapped candy bar, which could be easily shipped by a parent, wife or sweetheart in America to a loved one fighting in Europe. Hundreds of candy bar manufacturers sprang up in cities across the nation. To make matters worse, in 1917 the Confectioners Union in Chicago went on strike, shutting down every candy factory in the city. Menasha soon found itself stuck with an entire warehouse full of candy pails for which there were no longer any buyers. The candy pail business never did recover.

The market collapse accelerated. By the early 1920s, demand was dwindling for nearly every type of woodenware, and by 1930, the woodenware industry was barely a quarter of its former size — "a transition swift and deadly," in the words of Mowry Smith, Jr. Once one of America's most successful companies, Menasha was now struggling for its very life.

A Large Manufacturing Complex On the Fox River

Looking back today, woodenware was a great business while it lasted. Menasha occupied a large site along the banks of the northern channel of the Fox River. Spread across that site were lumberyards, saw mills, drying kilns, manufacturing plants, warehouses, office buildings and other facilities, including a car barn (where Menasha maintained its fleet of box cars) and a horse farm.

Because company facilities were located on both sides of the river channel, the company had its own bridge, used by office boys riding back and forth on bicycle to deliver messages, as well as by other employees.

In winter, timber was stacked along neighboring streets. In spring, when the river thawed, the timber was dumped into the river and fed into the main plant

through sluices along its water line. During those spring thaws, as the river level rose, water and fish alike sometimes surged through the sluices, flooding the plant's basement. Walter "Wally" Stommel, who joined the company in 1928, still recalls the time he and another employee, Tom "Tucker" Russell, fished in the basement and caught a northern pike.

The corporate office was located in a one-story building constructed in 1885. The office originally had two front doors. Back in the days when wages were paid in cash, the men from the plant would enter through the left-hand door, go to a cashier's window, pick up their pay envelopes and verify that the amount was correct. They would then proceed to a second cashier's window, where a clerk would count the pay again. Having been paid, and the amount having been double-checked, the men would exit through the right-hand door.

The basement of the corporate office building originally housed a large "vault" — in fact, no more than a locked room with windows — where all records and cash on hand were kept. Since at one time all records were hand written, this area was quite large to accommodate all the files. It was not until 1920 that Menasha acquired its first real vault.

The Vital Role of a Skilled Workforce

One of Menasha's tremendous competitive strengths was its many skilled employees and their work ethic. Entire families worked at Menasha, and employees often stayed 40 or 50 years.

In Charles Smith's time, most production was organized into crews of about four to six men. As an incentive to work hard, crews were paid by the unit produced rather than by the hour. Henry Suess

By 1955, as demand for woodenware collapsed, the pail factory was a small operation struggling to survive. Rather than fire any employees, the company kept operating the factory to provide jobs, finally closing it in 1957 after all the men had retired. The coaster at far left highlights not only woodenware, but also the company's 1927 entry into the manufacture of corrugated boxes, today its largest business.

started in the 1920s as a "hooper" in the pail shop and still remembers being paid 45 cents per 100 pails. Wally Stommel recalls, "When the men got through with a day's work, say 600 candy pails, they went home regardless of how early it was. Even if the company wanted them to work overtime, they wouldn't. The men were productive, but they were also very independent-minded." Stommel joined Menasha as an office boy in 1928 and stayed 46 years, retiring as personnel manager of the Neenah plant.

The leader of each crew was absolute boss. He was paid by the company for the entire crew's work and he, in turn, paid the men out of that sum. However, the introduction of the federal income tax in 1913 forced an unexpected change. "Along came the income tax, and the company had to report all the men who earned over a certain amount," says Stommel, whose grandfather was a lathe boss (a crew chief in the pail factory). "All these lathe bosses were reported as earning $4,000 or $5,000 a year, which was unheard of in those days. Hell, the bank president didn't earn that much. So the federal government went after them for taxes. Sure, they earned it, but they paid most of it out to their men and never kept records. The government went after my grandfather for $5,200 of back taxes. It was quite a mess." The company eventually took care of the tax bills for the men, and thereafter it paid each employee directly to avoid further problems.

The work of the men was fascinating and highly specialized. Members of a pail-making crew, for instance, included the set-up man, who put the staves together to form the pail; the turner, who smoothed the pail's outer surface; and the lathe boss, who smoothed the inside (his work was considered to be the most skilled). The staves for making the pails were of varying width, and this added to the complexity of the process. "What I remember most was these fellows putting in that last stave," says Breadon H. "Keg" Kellogg, a longtime Menasha executive, now retired, who joined Menasha in the early 1950s from Fort Wayne Corrugated Paper Company. The space left for the final stave could be of any width (depending on the width of the staves already in place), and it was the responsibility of the set-up man to find a stave which fit precisely in that final slot. The ability to judge the width of that slot by eye was astonishing. "The fellow would look over a pile of staves, and many times he could just pick one out and slap it in," Kellogg recalls. "If he didn't see a stave that fit, he would take one and hand it to the joiner, signaling with his fingers how much it should be trimmed. It amazed me how fast the men could work."

Each pail-making crew also included a hooper. His job was to put steel bands around each pail to hold it together. When the nearly finished pail reached the hooper, it was temporarily held together by a heavy metal ring. Simultaneous with putting on the steel bands, the hooper had to knock off the metal ring, catch it and stack it. "It

was quite a trick," Stommel recalls. The hooper wore a special glove on his left hand to avoid injury as he caught the ring.

Mike Muntner began in 1932 in the Bottom Room, where pail bottoms were made. The bottoms were fabricated by gluing together flat pieces of wood into square shapes, and then cutting the squares into round bottoms. The work was so specialized that Muntner's only task was to help plane the squares to smooth their surface.

Work in the plant was hard and, in many cases, repetitive and tedious. Nonetheless, there was tremendous camaraderie among the men. Most were happy to have skilled jobs that paid a decent wage, working with colleagues they respected.

As the 1940 photograph at left suggests, there was great camaraderie among Menasha's management group. The management team worked hard, but had time for fun as well. Left to right are John Young, Mowry Smith, Sr., Donald Shepard, Sr., and Donald Turner, Sr. Seated is George Hinton. At right is Carlton Smith (in the suit jacket) and below right is his brother, Mowry, Sr. Carlton headed Menasha Wooden Ware Company, which owned a portfolio of stocks and bonds, while Mowry headed Menasha Wooden Ware Corporation, the operating company.

Four Bright Young Men

Another of Menasha's strengths was a group of four young friends, all college graduates, who joined the company in the period from 1916 to 1918. They were enthusiastic and bright, and they would one day lead the company. It was unusual, to say the least, for a company in a small Wisconsin city to have four college graduates — three Princeton, one Yale, all class of 1916 — among its managers. Then again, Menasha was an unusual business.

The four were Mowry Smith, Sr., and Carlton Smith and their friends Donald Shepard, Sr., and Donald Turner, Sr.

As mentioned earlier, Willis Miner was president from 1921 to 1935, and he was in charge. Everyone, including the four young men, called him "Mr. Miner." No one dared do otherwise. However, it was the four who provided much of the energy and many of the new ideas that got Menasha moving again after the woodenware market had gone into a free fall.

Mowry and Carlton met Donald Shepard while attending prep school in New Hampshire. When the Smiths enrolled at Princeton, Shepard headed off to Yale, but they kept in touch. Graduating in 1916, Shepard served in the armed forces in Europe, worked at a bank in New York and then joined Menasha in 1918 at Mowry's and Carlton's urging. In 1921, he married their sister, Sylvia Smith, becoming a member of the family.

At Princeton, Mowry and Carlton met

Donald Turner, a native of Corning, Iowa, where his father was a banker and farmer. The three became friends and roomed together in an apartment over a store. After graduating in 1916, Turner worked for a Chicago bank and served in the armed forces before accepting a job with Menasha. He became like family.

Each of the four carved out his own niche at Menasha.

Mowry, Sr., was active in manufacturing and sales and, with his gregarious personality, had a wonderful way of relating to the men in the factory. He knew them all by first name and was liked and respected by even the toughest blue-collar worker.

Carlton was quiet, reserved and

Donald Turner, Sr., far right in this photograph, stands with his relatives, as well as Carlton Smith and his wife, Theda, to his immediate right. Turner had roomed with Carlton and his brother, Mowry, Sr., at Princeton University. After graduating in 1916, Turner worked for a Chicago bank and served in the armed forces before joining Menasha at the urging of Carlton and Mowry. Below is Donald Shepard, Sr., who knew Carlton and Mowry from prep school. He, too, was recruited to Menasha by Carlton and Mowry. The four close friends — Mowry and Carlton Smith, Don Turner and Don Shepard — led Menasha from the mid-1930s through the 1950s.

intellectual and was drawn to the investment side of the business. In 1926, Menasha was restructured into two separate companies: Menasha Wooden Ware Corporation, which continued the manufacturing and marketing operations and was Mowry's bailiwick, and Menasha Wooden Ware Company, which owned a portfolio of stocks and was Carlton's area. This arrangement gave each brother his own domain — and, in the process, avoided any potential disputes about which one would hold the top position, since each could hold a top position. Equally important, the portfolio generated investment income for Menasha shareholders even when manufacturing operations were unprofitable. Indeed, for many years the portfolio was the shareholders' major source of dividends. This

OUT OF WOODENWARE, INTO CORRUGATED

structure of two separate companies — one for manufacturing and marketing, the other to manage a portfolio of stocks that generated income — continued for 55 years. The two companies were finally reunited in 1981, forming today's Menasha Corporation.

Shepard and Turner learned Menasha's business from the bottom up, starting in the plant at 17 cents an hour — quite an expe-

Donald Shepard, Sr. (far right), Donald Turner, Sr. (second from right) and their friends, including Chester Shepard (far left), Donald Shepard's brother, enjoyed golf, tennis, sailing and other outdoor activities.

rience for a couple of Ivy League grads. After 1926, when Menasha was split into two companies, they were active primarily in the Corporation — that is, the manufacturing and marketing company.

Shepard took charge of advertising, sales and product development. He understood the needs of the consumer and developed products to meet those needs, including the Sitka spruce butter tubs that helped

In 1926, women load Sitka spruce staves at a company facility in Tacoma. From Tacoma, the staves were shipped to company assembly plants across the United States to make butter tubs. The assembly plants were located near major customers. One, for instance, was next to the Land O'Lakes Creamery Association in Minnesota. Although most of Menasha's production workers at that time were men, this photo reminds us that women worked in its factories as well.

save Menasha from bankruptcy in the 1920s. "My dad was the best salesman of that crew," says Donald "Tad" Shepard, Jr. "He could sell himself. He could sell the product. And he loved to write. He wrote many of the ads himself, and he wrote poems in his spare time." Shepard's advertisements presented a simple, dramatic message. One ad showed a stack of butter tubs reaching into the clouds, with the tag line: "If Menasha's Annual Production of Butter Tubs Was Stacked on Your Creamery Platform It Would Tower 1018 Miles Above You."

Turner was drawn to cost accounting, establishing a cost system that was one of the first of its kind in American industry. Don Turner, Jr., says, "My dad wasn't an accountant, but he was quite mathematical. He used cost accounting to run a better business by reorganizing the activities on the shop floor." Turner later became active in managing Menasha's West Coast timberlands.

The four were great pals, hanging out together in their free time as well as working in the business. "They went boating together, they played golf together, they played tennis together, you name it," Don Turner, Jr., notes.

How Menasha Survived and Changed

These four young men joined a business that was already faltering. As the 1920s began, facing the rapid collapse of its markets, Menasha survived initially by developing an entirely new product line — butter tubs made of Sitka spruce from the forests of the Pacific Northwest. Even though demand for most types of woodenware was falling, butter tubs were an exception: for most of the 1920s, butter was one of the few products that continued to be sold primarily in wooden packaging. Most butter tubs were made of ash wood, and the ash had spores that soaked into the butter. As a result, the butter became moldy within six months. Menasha's Sitka spruce tubs kept the butter fresh for up to a year.

Menasha formed a subsidiary, Northwestern Wooden Ware Company in Tacoma, Washington, to make staves from Sitka spruce. It also established a network of 14 assembly plants across the United States.

OUT OF WOODENWARE, INTO CORRUGATED

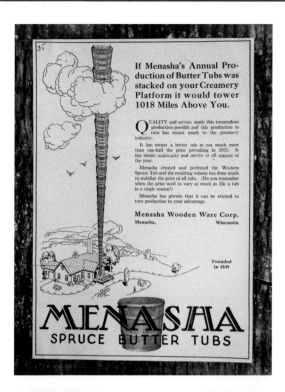

Menasha was known throughout the 1920s, 1930s and 1940s for its well-written advertisements that described its products in simple, compelling language. These ads were created by Donald Shepard, Sr., who spent his career with the company and was married to Sylvia Smith Shepard, the sister of Mowry Smith, Sr., and Carlton Smith.

The staves were shipped from Tacoma to these plants, which made butter tubs for local markets. Menasha quickly became the nation's number-one butter tub manufacturer.

By the late 1920s, however, even butter tubs were on the wane in favor of individually wrapped butter. Butter tubs had nonetheless done their job of keeping Menasha out of bankruptcy. The company continued to make the tubs until 1942, when the government froze the supply of Sitka spruce to assure its availability for military gliders and ship keels during the war. No longer able to use the superior Sitka spruce for butter tubs, and with demand for the tubs dwindling anyway, the company closed its butter tub business.

Menasha manufactured a variety of other wooden products in a desperate

Above is the company's Sitka spruce stave mill in Tacoma. Menasha produced Sitka spruce butter tubs from 1922 until 1942, when the government froze the supply of Sitka spruce to assure its availability for military gliders and ship keels. Below, wood "flour" was made by grinding dried spruce shavings. It was sold to manufacturers of explosives, plastic wood and other products.

attempt to generate revenue and keep the men at work. This included everything from wooden toys and children's furniture to thread spools and chopping blocks. For a time in the 1930s, the company even made fly swatter handles.

In 1929, Menasha developed an entirely new business when it opened a plant in Tacoma, Washington, to make "wood flour" — a fine powder produced by grinding and sifting dried spruce shavings. This business grew out of a chance encounter. Menasha had been burning the shavings, which were considered a useless by-product. However, as Mowry Smith, Sr., later recalled, "A very mysterious guy, Ray Gamble, came in one day and offered to buy them for $1 a load." On investigation, it turned out Gamble was using the shavings to make wood flour in his barn for sale to manufacturers of plastic wood, explosives and other products. Mowry Smith, Sr., quickly negotiated a deal with Gamble to open a plant, and the facility generated a nice profit for many years. In fact, the esoteric business of making wood flour proved to be such a winner that the company subsequently opened additional wood flour plants at North Bend, Oregon, in 1955; near Albany, Oregon, in 1963; and at Grants Pass, Oregon, in 1969. At their peak in the late 1960s, these plants produced more than 600 tons of the material per month.

A Historic Change: Entering the Corrugated Business

Despite these successes, finding catch-as-catch-can solutions was not the answer to the company's problems. Willis Miner, Mowry Smith, Sr., and all the other members of the management team recognized that Menasha needed to manufacture new products of new materials to survive long term.

The first big change came in 1927, when Menasha entered the corrugated box business. Corrugated gave Menasha hope for the future. Without corrugated, Menasha had no future.

Corrugated board had been developed in the 1870s when the machine that produces corrugating medium (the squiggly center of the board) was invented. By the 1920s, the use of corrugated containers by shippers of consumer and industrial products was increasing rapidly because of the containers' strength, impact resistance and light weight, and also because corrugated containers — unlike wooden containers — could be broken down and stored flat.

Fort Wayne Corrugated Paper Company, founded in 1908, and Container Corporation of America, established about the same time, were already major producers of corrugated boxes. Many of today's other leading box companies, including International Paper Company, entered the business in the 1920s. Many of these companies came from a background in paper manufacture, so they had the know-how to make and market corrugated. Menasha, by contrast, knew nothing about paper, so its transition into corrugated was challenging and remarkable: it had to start from scratch without any experience in paper-related products, competing against companies with a wealth of knowledge.

As mentioned earlier,

Menasha's employee publication was originally called From Logs to Candy Pails. *However, by the time of this issue in June 15, 1921, the name had been changed to the* Wooden Ware Log. *The quarterly publication continues today as the* Log.

Menasha first thought about producing corrugated boxes in 1917-1918 when Mowry Smith, Sr., proposed the idea and F. D. Lake rejected it (and was pushed out the door for his reluctance). Nothing further came of the matter until 1927, when Don Turner attended the showing of a film by Samuel M. Langston Co., a producer of paper manufacturing equipment, delineating the steps in making corrugated boxes. Turner was impressed and invited Langston officials to Menasha to show the film to the company's directors. The directors, too, liked what they saw. They recognized also

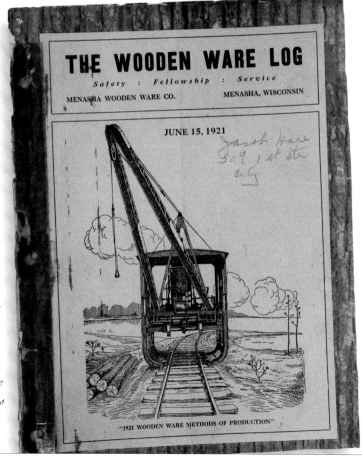

that Menasha had empty buildings that needed to be put to use. Consequently, they voted to set up a corrugated box plant in an empty four-story butter tub warehouse.

The new plant began operation in December 1927. Corrugating medium was purchased from local paper mills, while linerboard (which forms the smooth outer layers of corrugated board) was purchased from John Strange Paper Company, located just down the street from Menasha's facilities. The Menasha plant combined these materials to make corrugated sheets, then manufactured corrugated containers from those sheets.

Prior to the opening of the plant, not a single person in Menasha's middle or upper management knew anything about the manufacture or marketing of corrugated boxes. Rather than hire from the outside, Menasha gave its employees a chance to learn the business. While this was a daring decision, since the success of the operation would depend entirely on inexperienced people, the company's senior executives felt that providing steady employment was part of their mission. One of those who grabbed at the opportunity was George Hinton, a butter tub salesman who had been with Menasha since 1922. He became the fledgling plant's sales manager and, learning quickly, was appointed its general manager just two years later. He headed the facility until 1954, when he was transferred to California to start up, staff and operate a huge new corrugated box plant

Although Menasha was initially a very small participant in the corrugated box business, it expanded its presence in 1955 with the completion of this plant in Anaheim, California. The Anaheim facility was one of the largest corrugated box plants in the United States and was built by Menasha despite opposition from competitors.

built by Menasha in Anaheim. Along the way, he became a mentor to many Menasha employees, teaching them about corrugated. Hinton retired in 1964 after a long and remarkable career.

Today, corrugated containers are a huge national and global industry. According to the American Forest & Paper Association, 95 percent of all products in the United States are shipped from factory to market in these familiar brown containers. Moreover, the Menasha Packaging Group (corrugated containers) is Menasha Corporation's largest business, generating about $300 million of annual revenues.

This is quite a change from those early years when corrugated containers were a relatively new industry and Menasha's box plant was run by a butter tub salesman — albeit a very smart butter tub salesman — who had no previous experience in any kind of paper product.

The 1934 Strike

Following the opening of the box plant in late 1927, the corrugated business started slowly for Menasha, losing $33,000 in 1928. By 1933, however, corrugated sales were $619,000 and operating income was $51,000. That same year, the woodenware business lost $76,000. Although many employees had worried that entering the corrugated box business might jeopardize their jobs, the opposite was, in fact, proving true. As woodenware sales shrank, rather than laying off employees, Menasha was able to transfer employees from the woodenware manufacturing facilities to the corrugated box plant.

Nonetheless, this turn of events led in a roundabout way to a strike. By 1933, labor turmoil was escalating nationwide because of the Depression. That same year, President Roosevelt signed the National Industrial Recovery Act, guaranteeing labor the right to organize and bargain collectively. Lengthy, bitter strikes ensued against General Motors, the textile industry and numerous other companies and industries.

Menasha was no exception. In 1934, the American Federation of Labor sought to organize the company's employees, and it called a strike and blocked access to Menasha's production facilities. The union demanded higher wages and benefits. In addition, it proposed that seniority be determined separately by department rather than on a companywide basis. This is where the box plant came in. Mowry Smith, Sr., argued for companywide seniority, pointing out that, otherwise, longtime employees transferring from woodenware production to the box plant would have to start at the bottom of the seniority list all over again. The union sided with newer employees who favored departmental seniority.

The strike had its unusual aspects. The office manager at the corrugated plant was a fellow named Ed Fox. Years later, he recalled, "The union had pickets posted at both gates and prevented most personnel from entering the premises. As it was unlawful to picket a private home, the corrugated office was moved to my home." From that residential nerve center, box production was farmed out to competitors so that customers' orders could be met.

Six desks were installed in the home of another employee, Ralph Suess, and that became the temporary office of the Woodenware Division.

The strike lasted several weeks and was finally settled on June 30, 1934. As part of the settlement, departmental seniority, not companywide seniority, became the official policy.

Getting Back in the Profit Column

Menasha struggled financially throughout the early years of the Depression, finally returning to profitability in 1935. That profit was due entirely to the corrugated business, which increased its earnings to $112,000 in 1935 even as the Woodenware Division continued to sink, losing $49,000. The wisdom of having entered the corrugated box business could no longer be in doubt. The most ambitious and important transition in Menasha's history — out of woodenware, into corrugated — was clearly taking hold.

It was also in 1935 that Willis Miner died at age 72 after having led the company through a period of incredible challenge and change. Looking to a brighter future, the four "young men" — Mowry and Carlton Smith and their friends Don Shepard and Don Turner, seasoned executives now in their 40s — were ready to take charge.

AN ODYSSEY OF FIVE GENERATIONS

CHAPTER FIVE

The Third Generation

It was now up to the third generation of Smiths — Mowry, Sr., 45 years old, and Carlton, 44, grandsons of the founder — to lead Menasha out of the depths of the Depression. Mowry, Sr., became president of Menasha Wooden Ware Corporation (the operating company) and Carlton became president of Menasha Wooden Ware Company (which owned the portfolio of stocks). Their brother-in-law, Don Shepard, and their friend, Don Turner, continued as vice presidents of the operating company.

Of that group, Mowry was the boss — and an unusual one, at that. He was, quite simply, one of the most colorful and well-liked individuals in the history of the company. "My father loved people, and they reciprocated that love," his son, Mowry Smith, Jr., says. In an era when it was customary for company presidents to be addressed as "Mr." or "sir" by all but their closest associates, Mowry, Sr., resisted formality, creating an environment where people felt comfortable calling him by his first name. Rare was the individual who addressed him as "Mr. Smith."

This informality was a striking contrast to the style of Willis Miner and Charles Smith. In other ways, however, Mowry, Sr.'s management approach was similar to that of his father. Like Charles, Mowry was a hands-on manager who controlled all major decisions — initially with the help of Don Turner and later relying heavily on Dick Johnson, who succeeded Mowry as president and chief executive officer. The board of directors had virtually no voice in decision-making during his 25-year tenure, as was true during Charles's presidency. Of course, many family companies have strong chief executives who maintain tight personal control. In recent years, Menasha has changed so that its board now holds an active oversight role, similar to the governance of a publicly owned company.

Knowing Each Employee by Name

Managing the family business was, for Mowry, Sr., not just an obligation that fell to him as Charles's oldest son. He relished

Charles Smith's son, Mowry Smith, Sr., above, led the company from 1936 to 1961. At left, Mowry, Sr., leans across the desk of Ed Kelly, foreman of the pail factory. The picture was taken in 1926, two years before Kelly retired. Kelly joined Menasha in 1861 at age 10, at first piling staves, and spent 67 years with the company.

the job and loved every minute of it.

Mowry paid special attention to the manufacturing operations, which were the heart of the business. He went out into the factory every day to chat with workers. He also took the unusual step of maintaining a file card on each Menasha employee. That card included the employee's name, job, photograph and family information. In this way, Mowry, Sr., familiarized himself with all the people who worked for the company. It was possible to do this until the early 1950s because the company still had only a few hundred employees, located mainly in the city of Menasha. Walking through the box plant or between buildings or in town, he greeted each employee he encountered by first name. New employees were invariably amazed that the company president knew who they were.

If an employee had a problem, Mowry was eager to help. His office door was open to everyone in the company. One of his passions was to chop firewood for exercise. On learning that an employee didn't have enough money to buy firewood in the winter, he might chop a bundle and send it anonymously (although who could possibly *not* know it was from Mowry?). Or he might send food, if that was needed.

Mowry, Sr., maintained balance in his life. He believed that having a good time was just as important as work. He arose early each morning to get to the plant, returning home for lunch. In winter, if the

weather was nice, he sometimes donned old clothes and went ice-boating before heading back to the office in the afternoon.

He had an inherent faith in people. Katharine "Kig" Gansner, his granddaughter, lived with Mowry and his wife (her grandmother), Katharine, when she was a teenager. She was suspended from boarding school for two weeks during her senior year and had to return home to face her grandfather. "I walked into the living room and he got up from his chair and put his arms around me and gave me a big hug," she recalls. "The next day at lunch, he said, 'You know, some people are really lucky. They make their mistakes early in life and can learn from them.'

That's the only thing he ever said." Gansner is today a Madison, Wisconsin, attorney, mother of three and member of Menasha Corporation's board of directors.

Mowry, Sr., was such a colorful character, and there are so many wonderful stories about him, it is hard to know where to stop. One story involves an employee in the Menasha box plant who goofed off repeatedly and was always on the verge of being fired. Divorced and with custody of his three children, the employee was also a lackadaisical father who did not always feed or clothe the children properly. "One day, representatives of the state welfare agency showed up at the factory to serve the fellow with papers stating they were

THE THIRD GENERATION

Far left is a vintage photo of a corrugator, which makes corrugated board from two outer layers of linerboard and a center of corrugating medium. At left is the entrance hallway of the one-story building that was Menasha's headquarters from 1885 to 1967. An article in the company magazine said, "The old office was a friendly place.... high, narrow windows... huge, hardwood, rolltop desks, beautifully made and beautifully kept."

taking the children and making them wards of the state," according to Wally Stommel. Mowry, Sr., heard the ruckus and went onto the factory floor to find out was happening. Quickly realizing the gravity of the situation, and believing the children would be better off staying with their father, he worked out a deal for money to be withheld each week from the employee's paycheck and turned over to the welfare agency to buy food and clothing for the children. As part of the deal, the youngsters got to stay with their father — which is exactly what the father and the children wanted. Despite Mowry's kindness, the employee continued to malinger and ignore company rules. "He broke every rule in the book," Stommel says. "But every time his supervisor wanted to fire him, Mowry would come out and say, 'No, put him back to work,' because Mowry was concerned about the kids. That was just the way Mowry was. He took care of his employees and their families. He was great with people."

The Corrugated Box Plant

Beginning in the late 1920s, as the woodenware business neared the end of the line, Menasha focused increasingly on the production of corrugated boxes. The company's original corrugated box plant, which went into production in 1927 and remained in operation until 1964 when it was destroyed by fire, was located within Menasha's manufacturing complex along the north channel of the Fox River. It was a rather odd-looking structure — four stories high, long and narrow, sheathed in tin, with an elevator at one end. When a brick extension was added, a second elevator was installed. In-process products were moved from floor to floor by elevator on hand-pulled trucks, an inefficient procedure that would cripple a business today.

Much of the work in the plant was highly skilled, and nearly all of it was physically taxing, utilizing little of the automation that is now commonplace. When Frank Albert graduated from college in 1951, he intended to become a

AN ODYSSEY OF FIVE GENERATIONS

In 1928, Menasha and several other local companies founded Wisconsin Container Corporation to make solid fiber laminated paperboard, which competed at that time with corrugated board in the manufacture of shipping boxes. Laminated paperboard contains anywhere from two to five plies of board, depending on the thickness and strength desired, bonded into a single sheet. Menasha acquired full ownership of Wisconsin Container in 1969. Today, the unit is Menasha's Solid Fibre Division.

teacher, but he needed money, so he applied for a job at Menasha, which hired him as an "off-bearer." His job was to catch corrugated sheets as they came off a machine and stack them by hand on a truck. His first day on the job, one of the other workers shouted out, "Hey, this place is really getting class. Now they have college grads catching the sheets!" Albert soon moved into production management and spent his career with Menasha.

There was great camaraderie among the men, who respected each other and were happy to have steady jobs that paid well, especially during the Depression. Adding to the esprit was the plant cafeteria on the third floor, run by Mary Kass, the wife of Al Kass, a union official. The food was so good some employees, on completing their shift, stayed for dinner.

Many longtime Menasha employees who are now retired still look back fondly to their days at the old corrugated box plant and the people they knew and some of the unusual incidents that occurred.

In 1935, Mike Muntner was transferred from the bottom department (making pail bottoms) to the night shift at the box plant, one of many employees who made the transition from woodenware to corrugated. Reporting to his new job for the first time, he was assigned to bale waste paper for recycling. He was told briefly how to operate the baler, but was not told the maximum weight per bale, 700 pounds — an instructional shortcoming that soon became apparent.

This early advertisement for Menasha corrugated containers has all the distinctive elements of Menasha advertising of the period: direct, informative text; bold, simple graphics; and a printed border that looks like tree bark.

"That first night was very slow," Muntner says, "because the corrugator was running a huge order for Kimberly-Clark, and there was very little trim being cut from the sides. I made two bales, 990 pounds each." He weighed each bale and marked its weight on the side with his initials, as instructed. A week later, all the accumulated bales were loaded into a railroad boxcar for shipment to a recycling mill. "There were five of us," he says, "and we filled the bottom of the boxcar and then had to lift the rest of the bales, including my two, on top of the others. We were lifting and shoving and trying to get one of those bales on top, when this one fellow looked at the weight, 990 pounds, and tried to read my initials. He said, 'If I ever catch the SOB who made this bale I'll kick him right out the [boxcar] door into the river.' I was standing next to him and I could read those initials. But he couldn't figure them out, and I didn't say a word."

Expanding in the Corrugated Box Business

Making corrugated boxes is relatively straightforward, though technically complex. It begins with three huge rolls of paper: two of linerboard which form the outer surfaces of the corrugated board and one of corrugating medium which forms the wavy center. These rolls are fed into a massive machine called a corrugator, which aligns and glues them to produce corrugated board and then slits the board

AN ODYSSEY OF FIVE GENERATIONS

THE THIRD GENERATION

Trucks line up outside the Otsego, Michigan, paper mill. Menasha purchased a 60 percent interest in the mill in 1939, acquiring full ownership 16 years later. Far left is an aerial view of the mill, located on the Kalamazoo River. Near left is a vintage photo of mill employees. The mill continues today as a core Menasha Corporation facility.

into box-size pieces. The pieces are then scored, slotted, printed, die cut and joined to make the boxes.

The linerboard contains a preponderance of long fiber from softwood, primarily pine, giving the corrugated board strength and tear resistance. The corrugating medium, in the center, contains a preponderance of short fiber from hardwood, giving the board stiffness and structural integrity. It is this unusual combination of properties that makes corrugated such a popular material for shipping containers and other applications where light weight, durability and affordable cost are required.

In its early years of operation, the Menasha plant made corrugated boxes from recycled linerboard, known in the industry as "jute," and recycled corrugating medium, known as "bogus." Recycled simply means the product was made from old corrugated containers, newspapers and other waste paper, rather than from trees. Corrugating medium, whether recycled or virgin, is sometimes referred to as "nine point" because its thickness is typically about .009 inch.

The company was at that time a tiny factor in the corrugated box business, with less than a one percent market share, competing against giants such as Container Corporation of America and International Paper Company. Moreover, since it did not then own a paper mill, it had to purchase corrugating medium and linerboard from others. (Today it owns a corrugating medium mill but not a liner-

board mill. It acquires linerboard through a sophisticated program of purchases and trades.)

How the Otsego Mill Helped Change Menasha

Throughout the 1930s, Mowry Smith, Sr., sought to integrate "upstream" by acquiring a mill. However, Menasha was only marginally profitable, and he couldn't find one the company could afford. A golden opportunity finally arose in 1939, and Mowry, Sr., jumped at it, acquiring a 60 percent interest in the Otsego Falls Paper Mill on the Kalamazoo River in Otsego, Michigan. The mill had been built five decades earlier by a papermaker named George Bardeen and was once the largest paper mill in Michigan. According to an article in the *Allegan County News & Gazette*, "George E. Bardeen and Company had been searching the Kalamazoo Valley and Allegan areas for the ideal place to locate the mill in 1887 when unusually heavy rains flooded many of their prospective mill sites. Otsego offered the only high-and-dry site and provided convenient access to the railroad and electricity. It was there that Bardeen erected the town's first paper mill in the autumn of 1887 and began operations on Christmas Day."

The mill prospered for many years, manufacturing a variety of paper products, but fell into receivership during the 1930s and was acquired by David Greene, a Menasha resident who was manager of Wisconsin Tissue Mills. Greene converted the mill to the production of corrugating medium made from straw, which was then inexpensive and readily available. (Paper can be made from virtually any plant fiber, whether that fiber comes from wood, flax, cotton, sugarcane, esparto grass, straw, etc. The ancient Chinese, who invented paper, made it from tree bark and hemp. It is only since the invention of the wood pulping process in 1844 that trees have been the primary raw material for paper.) However, Greene struggled financially with the mill and enlisted Menasha as a partner, selling Menasha a majority stake while retaining 40 percent ownership for himself. He continued as the Otsego mill's general manager for the next 15 years, retiring in 1954 and selling his 40 percent interest to Menasha the following year.

After World War II, straw became scarce, forcing the Otsego mill to take its first major steps toward modern papermaking technology, including the conversion to wood chips as a primary raw material. In effect, the shortage of straw saved the mill from obsolescence by causing it to begin producing corrugating medium from hardwood, including oak, maple, birch and poplar. Since then, the mill has been expanded and modernized repeatedly and is today Menasha Corporation's single largest manufacturing facility, forming the Paperboard Division. In an unusual twist, the mill now produces corrugating medium not only from wood, but also from waste paper, harking back to days of old. Recycled corrugating medium has made a strong comeback in recent years, driven by environmental considerations and improvements in production technology.

Mowry Smith, Sr., loved people and was always ready to help an employee in need. He resisted formality, creating an environment where people felt comfortable calling him by his first name. Rare was the individual who addressed him as "Mr. Smith."

Menasha During World War II And the Purchase of John Strange

Menasha's 1939 investment in the Otsego mill was followed by the 1940 purchase of the Durham Container Company, a corrugated box manufacturer in North Carolina.

With the outbreak of World War II, Menasha began making corrugated boxes for the armed forces, which used them to transport ammunition, beverages and other supplies. Even the woodenware business enjoyed a slight revival during the war. Because of metal shortages, a paint manufacturer asked Menasha if it could supply wooden paint tubs. Menasha said yes, but not without an initial snafu: the turpentine in the paint reacted with the resin in the wood, causing the paint to discolor. The problem was resolved by coating the tubs with sizing material which sealed the wood. In 1942 alone, Menasha sold more than five million paint tubs to such companies as Pittsburgh Paint and Sears Roebuck.

It was also during World War II that Mowry Smith, Sr., turned his attention to the acquisition of a linerboard mill to complement the Otsego corrugating medium mill. This became critical as wartime production controls dried up the supply of linerboard available for purchase on the open market. Finally, in 1945, after months of negotiations and just as the war was ending, Menasha and four other local box companies jointly acquired John Strange Paper Company, which operated a linerboard mill built in 1888 on property adjacent to the Menasha box plant.

In 1969, Menasha acquired full ownership of John Strange, whereupon the mill was renamed the Menasha Paperboard Mill. The acquisition included a John Strange subsidiary, Appleton Manufacturing Company, which made equipment for the paper industry; a paper spiral-wound core plant, which was sold in 1973; as well as Strange's majority interest in Wisconsin Container Corporation, which later became Menasha's Solid Fibre Division.

Ironically, the Menasha Paperboard Mill no longer concentrated on the production of linerboard by the time Menasha acquired full ownership. "Even before the acquisition, we had taken the mill out of linerboard into other products," says H. E. "Rusty" Sattler, who was vice president of sales at John Strange and later a Menasha Corporation vice president. These products included various types of paperboard for the manufacture of luggage, composite cans, outdoor advertising displays and folding cartons.

Menasha sold the Menasha Paperboard Mill in 1983, and it continues to be operated today by U.S. Paper Mills Co., located in DePere, Wisconsin.

Expanding to the West Coast

The competitive landscape of the corrugated container business changed dramatically following World War II, when companies such as International Paper moved south, building huge new mills that made linerboard from southern pine. The kraft linerboard from these mills was superior to the "jute" (recycled linerboard) then made at most older mills in the North — one reason why the John Strange mill stopped producing linerboard and got into other products. At the same time, demand for corrugated containers rose sharply, driven by the postwar economic boom.

Unable to afford a southern mill, Menasha missed out initially on the tremendous postwar expansion of the corrugated container market. Nonetheless, Mowry Smith, Sr., had a few tricks up his sleeve. Rather than erect a mill in the South, in 1953 he announced plans to build a state-of-the-art corrugated box plant in Anaheim, California, just outside Los Angeles. This was a striking change of direction for Menasha, which up to that time had produced corrugating medium and corrugated containers at older facilities in the Midwest.

A basic tenet of the box business is that mills are located near raw materials (whether those materials be trees in the South, Pacific Northwest or other regions, or waste paper from urban centers), while converting facilities — plants that convert the mills' products into boxes — are located near markets. Building a box plant is much less expensive than building a mill.

Mowry, Sr.'s choice of Anaheim for a box plant arose from a chance encounter with an official of a California telephone company. When Mowry said Menasha was interested in building a corrugated container plant to serve the fast-growing Southern California market, the official recommended Anaheim as an ideal site.

Mowry, Sr.'s granddaughter, Kig Gansner, was in second grade when he first went to Anaheim to look for land, and he decided to take her with him. "We were going to fly," she says, "but I talked him into going by train. We went to Chicago and then to Los Angeles, just the two of us on the train together. Then we went to Oregon to visit the company's timberlands. The entire trip lasted about two weeks. When my oldest kid hit second grade, I can remember thinking to myself, 'Why in the world would he have spent two weeks with a second grader with no breaks?' But he was like that. He loved kids. He was a great guy."

Mowry's decision to build a box plant in California, without consulting the Menasha board of directors, was bold to say the least. Tad Shepard points out that Container Corporation of America dominated the West Coast corrugated container market and did not welcome new competition. "Mowry went ahead despite tremendous pressure from the box companies on the West Coast not to do so," Shepard says.

Officially named Menasha Container of California, the new facility was the industry's largest box plant on the entire West Coast. It began operation in 1955 and quickly captured market share, despite repeated warnings from competitors that it would fail. Within 12 years, the Anaheim plant was manufacturing and marketing 10 percent of the corrugated containers used in the Los Angeles area. Customers included such well-known companies as Oscar Mayer, Kimberly-Clark, Packard Bell, Sunkist, Union Carbide, General Mills and Columbia Records.

While a great success in terms of capturing market share, the Anaheim facility never achieved the profitability expected of it. The facility was among those sold in 1981 to Weyerhaeuser Company, which continues to operate the facility today.

Besides being a major operation in its own right, most importantly the Anaheim box plant gave Menasha the leverage to build a corrugating medium mill in North Bend, Oregon, by providing that mill with a market for its product. The North Bend mill was completed in 1961 — but that is a story for the next chapter.

The Trading Program

In the mid-1950s, around the time of the completion of the Anaheim box plant, Menasha began to develop a sophisticated trading program to find a market for the output of its Otsego mill and to meet its own needs for corrugating medium and linerboard.

Few if any corrugated container manufacturers have a perfect match between the amount of linerboard and corrugating medium produced at their mills and the amount consumed by their box plants.

It is hard to imagine today how labor-intensive the manufacture of corrugated boxes was through the 1940s and even later. This employee is using a bundler to tie a stack of boxes.

Invariably, they trade with other companies to secure the linerboard and corrugating medium they need or swap the excess they have. This is similar to the way petroleum companies trade gasoline to balance differences between the production of their refineries and the demands of their retail service stations.

"Until the early 1950s, it was really quite a chore to find buyers for all the corrugating medium produced at the Otsego mill," Tad Shepard relates. "Mowry broke the ice by arranging our first trade with Weyerhaeuser." Shepard took over the program after that trade and was later succeeded by Rusty Sattler. The program continues to this day, involving an ongoing stream of swaps and trades with paper mills that have corrugating businesses across the United States.

Sattler, who ran the program from 1971 to 1988, explains, "The first thing you have to do is take the entire country and determine who produces linerboard and corrugating medium and who uses them. And after you understand that, you have to say, 'Where does Menasha fit in?'"

In the 1960s and 1970s, when it operated both the Otsego and North Bend mills, Menasha was a major supplier of corrugating medium, but did not produce much linerboard. "We found that Menasha had a good product in corrugating medium, and everybody wanted it," Sattler says. As a result, the company was able to trade corrugating medium for linerboard, and in that way meet the linerboard needs of its box plants. However, the trades became much more complex than simple one-on-one transactions. They often involved complex multi-party swaps to reduce freight costs. For instance, a box company in Minnesota might want corrugating medium. Rather than ship medium from its Otsego mill in Michigan, Menasha might arrange for the medium to be delivered from a mill in Minnesota, and then compensate the owner of the Minnesota mill by delivering corrugating medium from Otsego to a box plant in Michigan.

"We called those freight-saving trades," Sattler says. A specific goal was to minimize freight costs and make money on the trades by finding the best possible transactions for Menasha at any given time. "I don't play poker," Sattler adds, "but, I tell you, this was the next best thing to it."

A Brief Struggle: Fortunately, No Blood Drawn

Let's take a snapshot look at Menasha in the mid-1950s, just before sales began to escalate with the completion of the Anaheim box plant. Menasha was still a relatively small company, with annual revenues of less than $10 million. Moreover, its management and board of directors were both family-dominated. As of 1955, the board of directors of Menasha Wooden Ware Corporation (the operating corporation) consisted of nine members:

Mowry Smith, Sr., and his son Mowry Smith, Jr.;

Carlton Smith and his son Tamblin Smith;

Donald Shepard, Sr. (Mowry, Sr.'s and Carlton's brother-in-law) and his son Donald Shepard, Jr.;

D. L. Kimberly, Henry Smith's son-in-law, representing the Henry branch;

Donald Turner, Sr., the longtime Menasha executive who had been a Princeton classmate of Mowry, Sr., and Carlton; and

Robert M. Briggs, a senior company executive who, like Turner, was not related to the Smiths.

It was also in the early 1950s that a potential dispute arose between Mowry, Sr., and Carlton and their sister, Sylvia (Donald Shepard, Sr.'s wife) — the type of dispute that often tears families apart.

The background was this. When their aunt, Jane Smith (Elisha's daughter), died childless in 1941, she left all her Menasha stock to Sylvia and none to Mowry, Sr., or Carlton. As a result, Sylvia became the largest single Menasha shareholder, with 29 percent of the stock, whereas her brothers each owned 18 percent. It was an unusual situation: the brothers ran the business, yet Sylvia was the largest shareholder. Apparently concerned that they might be outvoted by her on some critical matter, Mowry, Sr., and Carlton privately formed a 10-year trust, agreeing to vote their Menasha shares jointly during the term of that trust. All three siblings lived near each other, and the families socialized and sometimes got together for big dinners on Thanksgiving and Christmas. There was no overt hostility among the three siblings. Apart from the issue of the voting trust, the family was close.

None of this was ever publicized, but it might well have gone haywire and resulted in a divisive family fight. Fortunately, this did not happen. Sylvia, on learning of the trust, did not say a word to her brothers. No critical vote came up during the 10 years, and at the end of that period Mowry, Sr., and Carlton simply let the trust expire.

Down-to-earth Midwestern sensibility had prevailed.

This episode reminds us of just how difficult it is to keep a family business together, and how remarkable it is that the Smiths have avoided serious disputes for a century and a half.

Whatever Happened to the Woodenware Business?

The woodenware business died with a whimper, not a bang.

Toy and juvenile furniture was one of the company's last wooden products. The company began manufacturing this furniture in 1935 in a desperate attempt to keep its woodenware facilities in Menasha going and maintain jobs. In 1938, furniture production was transferred to the company's stave mill in Ladysmith, Wisconsin.

When the Ladysmith facility was destroyed by fire in 1942, the company acquired a furniture factory in Rockford, Illinois, and transferred Ladysmith's production there. But it turned out that Menasha was not large enough in toy and juvenile furniture to compete effectively, and in 1952 it converted the Rockford plant to the manufacture of corrugated boxes. Several years later, Mowry Smith, Sr., commented, "We rented a plant at Rockford, Illinois, and then bought it. We made furniture there and it was going fine until after the war. Then we paid an industrial consulting firm $44,000 to tell us to get out of the furniture business, which, by that time, we knew anyway!"

By the mid-1940s, Menasha had closed most of its other woodenware operations, as well. There were no mass layoffs. Many of the men and women who had worked in the company's woodenware manufacturing operations had, by then, retired or transferred to the company's newer businesses, especially to corrugated.

Even though the market had withered, the company continued to manufacture pails. "From the late 1920s on, we produced limited quantities of woodenware not to make money but to keep people employed," Tad Shepard says. "When I joined the company in 1950, some of the old-timers were still making pails and having a grand time doing it."

By the mid-1950s, all these old-timers had retired or were about to retire. The time had finally arrived when Menasha could close its woodenware business without throwing anyone out of work. With demand for wooden pails all but dead, the remaining staves were used to make planters for sale to nurseries. "For maybe half a year, all we did in the pail plant was make flowerpots," says Wally Stommel, who was the facility's superintendent. "We used up all the staves and closed the pail business."

Menasha shipped its last flowerpot in 1957, forever ending the company's involvement in the manufacture of pails, tubs and other wooden packaging. It was a wonderful business that was once at the heart of American commerce, and continued to provide steady jobs more than 30 years after the woodenware market went into a tailspin. Today it is but a distant memory.

In an effort to keep its woodenware business going and its workers employed, in 1934 Menasha began manufacturing toy and juvenile furniture.

THE THIRD GENERATION

This 1956 photograph, titled "last of the pail makers," was taken just before Menasha closed its pail factory, ending its involvement in the woodenware business. Wally Stommel, the plant superintendent, is in the back row, far right. After the plant was closed, the men in this photograph either retired or were transferred to other jobs.

CHAPTER SIX

Trees: A Family Heritage and Company Tradition

In the 1950s, during the second half of Mowry Smith, Sr.'s presidency, Menasha refocused its sights on one of its most valuable and underutilized resources: its timberlands in the Pacific Northwest.

Charles Smith had acquired these lands, with their majestic stands of Douglas fir and other species of softwood and hardwood, in the early years of the twentieth century. He not only viewed them as a potential source of raw material for the woodenware business, but also recognized their superb long-term investment value. Indeed, compared with today's prices, timberland in the Pacific Northwest was then incredibly cheap.

However, following Charles's death in 1916, little was done to manage these lands or systematically capitalize on their worth. As the woodenware business faded and Menasha struggled to establish a position in the corrugated container industry, the company became a passive landowner in the Pacific Northwest, selling logging rights to

Menasha owns 100,000 acres of prime timberland in the Pacific Northwest. Opposite, in this vintage photo, Carlton Smith stands next to a felled Douglas fir to demonstrate its huge size.

others in response to inquiries. "A local logger might come across 40 acres of mature trees and want to harvest them," says William Lansing, president, Forest Products Group. "He would write to the company in Wisconsin and propose a deal, and the transaction would be worked out by mail."

Yet even as they just sat there, these properties kept getting more valuable as land prices in the Pacific Northwest rose. In the years following World War II, Mowry Smith, Sr., decided the time had come to manage the company's western timberlands actively and, moreover, to add to its holdings. If timberland was such a great investment, he reasoned, why not own more?

"Don Turner played a central role in this," Tad Shepard recalls. "Turner was a very bright guy. While he wasn't an accountant, he had a financial viewpoint and he made it his business to learn about the timberlands. He was key in analyzing the timberlands and adding to the company's holdings in the 1950s."

The sudden upswing of activity in the

AN ODYSSEY OF FIVE GENERATIONS

Menasha's West Coast timber activities are focused in the twin cities of North Bend and Coos Bay on the Oregon coast. From 1948 to 1967, the company owned a plywood mill at North Bend, providing an outlet for some of the company's timber.

1950s is demonstrated by the composition of Menasha's timber holdings today. Menasha currently owns approximately 100,000 acres of prime timberland in the Pacific Northwest, mainly in Oregon and Washington. Of that total, about 20,000 acres date from Charles's time, about 40,000 were acquired in the 1950s and the remaining 40,000 have been purchased since the 1950s.

Key Acquisitions in the Pacific Northwest

In an important first step to leverage the value of its timberlands, in 1948 Menasha invested $400,000 to acquire a one-half interest in a plywood mill under construction at North Bend on the Oregon coast and loaned $800,000 to the facility. It later purchased full ownership of the mill, which became a subsidiary, Menasha Plywood Corporation.

The facility, which made plywood for the home construction industry, became an outlet for some of Menasha's timber. Although the plant was never very profitable and was closed in 1967, it represented a fundamental change in direction for the company — the first time since the demise of the Sitka butter tub business that Menasha made direct use of its West Coast timber rather than merely selling logging rights to others.

Then in 1954, in a deal hammered out in less than 30 minutes, Menasha acquired Irwin and Lyons Lumber Co. Considered one of the top independent logging companies in the Coos Bay/North Bend region of Oregon, Irwin-Lyons was up for sale because of the untimely deaths of its two owners within weeks of each other in 1953. Irwin-Lyons's general manager, John Hawkins, initially offered the business to Weyerhaeuser Company for $3.5 million, take it or leave it. When Weyerhaeuser sought a lower price, Hawkins refused and made the same offer to Menasha. Even though Menasha was not a major forest products company, Hawkins knew

Menasha because of its timber properties in the region and its ownership of the North Bend plywood mill.

Mowry Smith, Sr., and Don Turner traveled to Oregon to meet with Hawkins. At that get-together, attended also by Mowry Smith, Jr., and Allen P. Stinchfield, a Menasha executive on the West Coast, Hawkins emphasized that the price was nonnegotiable. "He was very straightforward," Mowry, Jr., recalls. "He said, 'I want you to know this is exactly the same deal we offered to Weyerhaeuser and they turned it down.'"

Hawkins then left the room so the Menasha executives could discuss the proposal among themselves. According to Stinchfield, Mowry Smith, Sr., immediately said he not only wanted to buy the property but also believed Hawkins was serious about the price being firm. The discussion lasted just a few minutes, after which, according to Mowry Smith, Jr., "We got John back in the room and my father said, 'We'll take it.'" The deal was concluded as simply as that.

Although $3.5 million was a huge investment for Menasha at the time, Don Turner had an eye for timberland and recognized that the price was, in fact, a bargain — a point that quickly became apparent. The purchase included more than 20,000 acres of Oregon timberland owned by Irwin-Lyons. The average price paid by Menasha for the standing timber on that land was approximately $10 per thousand board feet, well below going market rates.

Within a year, as the market strengthened, Menasha was selling logs from the Irwin-Lyons properties at prices as high as $29 per thousand board feet. The $10 average acquisition cost and the $29 selling price were not directly comparable, because the $10 did not include the expense of harvesting trees and transporting them to market. Nonetheless, as Dow Beckham, who was superintendent of an Irwin-Lyons subsidiary, remarks, "Menasha got a very good deal."

Pleased with the acquisition, Mowry, Sr., and Don Turner began acquiring additional timberland in Oregon and Washington. In 1956, Menasha paid $2 million for 12,000 acres owned by Al Pierce Lumber Company. Not long thereafter, the company acquired Ball Lumber Company, which owned 1,600 acres in the region. Measured in relation to Menasha's sales and earnings at the time, the Irwin-Lyons, Al Pierce and Ball Lumber acquisitions were among the largest investments ever made by the company. These properties continue today as the core of Menasha's West Coast timber holdings.

Becoming Fully Integrated in the Timber Business

In acquiring Irwin-Lyons, Menasha received not only its timberlands but also its logging operations, something Menasha had previously lacked. These operations employed a team of experienced lumberjacks and included tractors, trucks and other equipment and a bunkhouse for 80 lumberjacks at a camp in the woods. Menasha was now able to log its own trees and cut a predetermined amount of timber each year, rather than being dependent on the whims of other loggers' bids.

"With the acquisition of Irwin-Lyons, Menasha found itself fully integrated for the first time in the timber business in the Pacific Northwest," Lansing says. "It owned the timberlands, had the capability to log them and owned a plywood mill that used the wood."

By the late 1950s, Menasha's timber operations were headed by Stinchfield, an important Menasha executive during the period from the 1950s through the early 1980s. Al Stinchfield grew up learning about the lumber industry because his family was in the business. Beginning his career as a certified public accountant, he joined the staff of the North Bend plywood mill in 1948 and became a Menasha employee the following year when Menasha acquired an interest in the mill. With his outgoing personality, intelligence and work ethic, he advanced quickly within Menasha, becoming vice president of the Land & Timber Division and a member of Menasha's board of directors. He was responsible for many of the land purchases made by Menasha in the 1960s and 1970s and for improving the company's forest management practices. In addition, he and his wife, Ellen, were active in numerous civic and charitable causes in the twin cities of Coos Bay and North Bend. In 1974, they received the Salvation Army

Well into the 1920s, the company maintained logging operations in the Midwest to support its woodenware business, while buying timberland on the West Coast as a long-term investment. This dual focus of Midwest and West Coast is highlighted by these two photographs. At right is a company logging camp office in the Midwest, perhaps in Wisconsin, in 1926. Far right is a logging operation of the Irwin and Lyons Lumber Co. in Oregon. Menasha greatly expanded its West Coast presence when it acquired Irwin and Lyons in 1954. The engine is mounted on a sled so it can be moved to provide power for logging operations anywhere in the forest.

"Others Award" for their volunteer efforts on behalf of local youth. Three years later, they were named outstanding citizens of the year by the Coos Bay Area Chamber of Commerce. When Stinchfield retired in 1983, Menasha's employee magazine, the *Log* (which began publication in 1920 as *From Logs to Candy Pails)*, noted, "He built the company image and properties to the top-drawer status Menasha now occupies in the Pacific Northwest."

The Origins of Menasha's Reforestation Program

Even into the 1950s, many timber companies did not replant the land they harvested. They let nature take its course and regenerate the forest on its own, a practice now considered to be environmentally unacceptable and economically wasteful.

But times were changing, and with the Irwin-Lyons acquisition Menasha finally had the people and equipment to begin a reforestation program. The first tract to be replanted was a 556-acre site in the Lowe Creek forest near North Bend, a tract that highlights the fundamental changes in Menasha's forest management practices over the years.

Charles Smith had acquired the Lowe Creek site early in the century as a long-term investment. Menasha did nothing with the land until 1927, when it sold logging rights to Moore Mill and Lumber Company. Logging was still in progress in 1936 when all the trees that had not yet been cut were destroyed by a fire, ushering in a fallow period for the site. Because Menasha was an absentee landowner not actively monitoring all its West Coast timber properties, after the fire local ranchers began using the Lowe Creek site as pasture for cattle and sheep, periodically scorching the earth to kill any seedlings that sprouted naturally. In 1952, a Menasha forester, noting the difficulty of growing trees on pasture, recommended that the land be sold because "there is little possibility that the area would be satisfactory... for the production of timber."

However, Stinchfield took charge of timber operations at that time, and he immediately halted all further land sales until each tract could be individually evaluated. Thus, the 556 acres at Lowe Creek were not sold. Indeed, when Menasha initiated its reforestation program in the late 1950s, those acres were chosen as the first site on the recommendation of forester Dick Hintz, who wrote, "I believe this can be one of the most rewarding projects of our entire reforestation program." Hintz and his crew planted two-year-old Douglas fir seedlings on the site in the winter of 1958-1959, beginning the renewal of the land.

Those seedlings took hold, just as Hintz had predicted. By 1994, they had grown to an average height of 80 feet. That year, a commercial thinning of smaller trees generated the first modest return: $400,000 from the sale of timber. But that was just the beginning. Even today, the trees at Lowe Creek continue to grow. By about the year 2020, the forest will reach full maturity and the trees will be harvested, whereupon new seedlings will be planted, beginning the cycle all over again.

Menasha's western timberlands, bought at inexpensive prices, are a gift to the current generations of Smith family shareholders from earlier generations. The objective of Menasha's forestry programs, among the best in the region, is to manage these lands wisely and maximize their long-term value for the benefit of shareholders and employees.

Menasha initiated its reforestation programs in the West in the late 1950s, planting seedlings to renew the land after trees are cut. Its reforestation program is today considered one of the best in the region.

Splash Dam Logging on the Coos River

The Irwin-Lyons acquisition also included its subsidiary, Coos River Boom Company, one of the most unusual operations ever owned by Menasha. The word "boom" in the company's name referred to a chain or raft used to control logs in water.

Coos River Boom Company held an exclusive franchise from the state of Oregon to drive logs down a 24-mile stretch of the South Fork of the Coos River, similar to the right-of-way owned by a railroad. At river's end, the logs would be gathered and taken to sawmills. The company not only transported its own logs but was also required by state law to transport logs for others, charging fees regulated by the state Public Utility Commission. Logs were branded so they could be sorted by owner at the end of the journey.

Coos River Boom Company had secured its franchise in 1936 and, as allowed by that franchise, built two "splash dams" on the river in the early 1940s. Logs were stored temporarily in an artificial reservoir behind each dam. After the water had built up to a sufficient level, the dam gates would be opened and the logs would careen down the river in the surging waters. "It was quite an event to see those logs come tumbling

TREES: A FAMILY HERITAGE AND COMPANY TRADITION

Left, log drives on the Coos River ended at tidewater, filling the river from bank to bank. Below is one of Menasha's two "splash dams" on the river. Menasha was the last company in the western United States to drive logs down a river, a colorful practice that is now a distant memory.

down the river," Lansing observes.

At one time, there were dozens of splash dam logging operations on rivers throughout the Pacific Northwest. The practice flourished before roads were built into the forest and trucks became a reliable means of transportation. By the 1950s, however, the glory days of splash dam logging were long past because of environmental concerns and damage caused to the logs as they crashed into rocks on their way down the river. Indeed, Coos River Boom Company was a relic — the last splash dam logging company still operating in the entire western United States. "Coos River Boom Company had survived because the government needed low-cost timber during World War II," says Beckham, who was the company's superintendent.

After the war, the company kept operating despite growing opposition from fishermen, farmers and environmentalists. This opposition centered around the destruction of fish habitat and other environmental harm caused by the logs. In addition, there was a growing recognition that splash dam logging involved an unacceptably high level of risk for the loggers themselves. "When logs jammed in the river, the loggers had to get onto them and break up the jam," Beckham says. Loggers who fell into the water could be injured or even crushed to death.

Menasha kept operating Coos River Boom Company for three years after the acquisition, using that period to build a road adjacent to the river so trees could be hauled out of the forest by truck. In October 1956, the Oregon legislature banned all use of rivers in the state for running logs, and the Public Utility Commission told Menasha to close the operation by June 1957. The final run of logs down the Coos River was such a historic event that *Life* magazine hired a local photographer to capture for its readers this last vestige of a colorful industry practice.

When that final run was completed, under instructions from the state Public Utility Commission, Menasha broke the dams and burned them, marking the end of splash dam logging in the Pacific Northwest.

The North Bend Paper Mill

Continuing its expansion in the Pacific Northwest, in the late 1950s Menasha began construction of a corrugating medium mill at North Bend. This mill was made possible by a combination of factors.

Most importantly, the Anaheim container plant provided a market for the mill's production. "Without sales, a mill is dead," Tad Shepard observes. "Anaheim provided the critical sales for the North Bend mill."

Also, with its purchase of Irwin-Lyons and other timberland holdings in the region, Menasha had an ample source of raw material for the mill. Still further, Menasha faced a wood waste disposal problem from its plywood and lumber mills in the Coos Bay region, a problem that building a paper plant in North Bend

TREES: A FAMILY HERITAGE AND COMPANY TRADITION

In 1962, Menasha employees Red Weybright and Al Nighswonger saw a tree into sections small enough to be transported by truck. Menasha had acquired its first logging operations in the West eight years earlier. The 1949 telegram, above, is addressed to Menasha Vice President Don Turner, who played a major role in overseeing the company's West Coast properties from the 1930s through the 1950s.

AN ODYSSEY OF FIVE GENERATIONS

This truckload of logs is headed for Menasha's North Bend, Oregon, paper mill. At right, a log is de-barked at the mill before being cut into small chips that will then be "pulped" and made into paper. Menasha built the mill from 1959 to 1961 and operated it for 20 years before selling it to Weyerhaeuser Company. Weyerhaeuser continues to operate the facility today.

could help solve. Although Menasha and other mills had traditionally disposed of tree bark and other wood waste by burning it in furnaces shaped like teepees, with smoke pouring out of the tops of the units, the state of Oregon was beginning to demand an end to the practice. A paper mill at North Bend could use mill waste as fuel, turning a by-product that was going to be costly to dispose of into fuel with economic value.

The final key to the puzzle was the discovery of a huge supply of fresh water by Pacific Power & Light Company beneath the North Bend sand dunes along the Pacific shore. Papermaking requires massive amounts of fresh water, and this supply beneath the dunes — believed to come from an underground river — offered a perfect source. Without fresh water, the mill could not have been built.

The site also made sense because Coos Bay is one of the great natural harbors on the Pacific Coast. Consequently, corrugating medium from the mill could be transported to market by ship.

Pacific Power & Light got the ball rolling when it discovered the water. Having made that discovery, it set out to convince a manufacturer to build a factory on the dunes, with the idea the factory would become a customer for Pacific Power & Light electricity.

At the invitation of Pacific Power & Light, Menasha studied the site. However, Menasha's initial projections showed that a paper mill at that location would not be sufficiently profitable to justify the investment. "Then Dick Johnson [Menasha's controller] reviewed our assumptions and realized they were flawed," Stinchfield recalled in an interview not long before his death in 1998. Those assumptions allowed for yearly increases in the cost of raw materials, but did not allow for rising paper product sales prices. "When we investigated the matter further and used the right numbers, it became clear the mill would be a good investment," according to Stinchfield.

Menasha broke ground for the mill in 1959, completing it two years later. To supervise its construction and manage the mill once it was completed, Menasha hired Ernest C. Manders, a brilliant if somewhat

AN ODYSSEY OF FIVE GENERATIONS

The North Bend, Oregon, paper mill, completed in 1961, at that time was one of the largest investment projects in Menasha Corporation's history. Construction of the mill was made possible by a combination of factors, including the fact that the Anaheim container plant — built six years earlier — provided a captive market for the mill's production.

unusual individual. Manders came to Menasha from Kimberly-Clark Corporation, where he had been in charge of the expansion of a paper mill in Alabama. "Ernie was a great guy, but a little different," says Tom Williscroft, who was hired by Manders in 1965 and succeeded him as general manager of the North Bend facility in 1971.

Colleagues loved to tease Manders about his idiosyncrasies. For instance, he kept sheep, peacocks, cattle and other creatures on the grounds of the mill in its early years of operation because he liked animals. He was also a relentless tinkerer who invented what he called "non-patented" devices, such as a log-splitting blade that was attached to the front of a fork lift. Manders managed the mill for its first decade and was later vice president in charge of Menasha's Paperboard Division, which encompassed the North Bend mill as well as the Otsego and John Strange facilities.

Menasha's investment in the North Bend mill was nearly $10 million. At the time, it was the largest investment the company had ever made, and a lot was riding on the mill's success. There was little margin for error.

The original plan was to make corrugating medium entirely from wood chips. "However," Williscroft says, "the sheets fell apart. There was no strength to them. So the company built a waste paper plant, and then used recycled corrugated for about a 10 percent to 15 percent mix to give the sheets the needed strength. That set a pattern for the mill. Thereafter, we would balance the price of wood chips and the price of recycled paper and use as much as 40 percent recycled paper if doing so provided the lowest raw material cost."

After these startup problems, typical of paper mills, the North Bend facility became one of the top-performing mills in the world. "Every year, we made more tons," Williscroft reports. By the late 1970s, the mill had the best corrugating medium-producing machine in the world, according to Beloit Corporation, a manufacturer of paper-making machines, as measured by tons of production per width of machine.

In addition to the mill being an important facility in its own right, its planning and construction signaled a management transition within Menasha. Mowry Smith, Sr., supported the idea of building the mill. However, implementation of the project was left mainly to Dick Johnson. When the mill was completed in 1961, Mowry, Sr., retired and Johnson — having demonstrated his ability to successfully manage a major project — succeeded him as president and chief executive officer.

The North Bend mill was a key Menasha facility for two decades, generating substantial profits. Today, however, it is no more than a memory for the company and its employees. In 1981, as we shall see in the next chapter, Menasha sold the mill to Weyerhaeuser Company for reasons unrelated to the mill's performance. The sale did not include any of Menasha's timberlands, which continue to be owned and managed by the company.

Although Weyerhaeuser operates the North Bend mill today, it will always be a wonderful part of Menasha's history.

Mowry Smith, Sr.'s Major Accomplishments

Mowry Smith, Sr., retired from day-to-day management in 1961, moving up to chairman. He died unexpectedly three years later at age 72, in robust health to the end.

In his 25 years as Menasha's president and CEO, Mowry, Sr., had:
- led the company out of the depths of the Great Depression;
- developed the company's corrugated container business;
- acquired the Otsego and John Strange mills;
- constructed major new facilities in Anaheim, California, and North Bend, Oregon;
- increased the company's holdings of timberland in the Pacific Northwest; and
- established a toehold in the plastics business with the 1955 acquisition of a majority interest in G. B. Lewis Company of Watertown, Wisconsin (Chapter Nine).

This remarkable man with a congenial personality, esteemed by all who knew him, understood what had to be done to save Menasha from bankruptcy following the collapse of the woodenware business — and he did it. Most other woodenware companies did not change with the times and no longer exist today. Menasha did change, thanks to Mowry, Sr.'s vision and persistence.

AN ODYSSEY OF FIVE GENERATIONS

CHAPTER SEVEN

Dick Johnson and the "Professionalization" of Menasha

The failure of his father, Charles Smith, to groom a successor must have left an indelible impression on Mowry Smith, Sr. Not wanting to repeat the same mistake, in 1955 he reached outside the family and recruited Dick Johnson, 39, Marathon Corporation's tax manager, to join Menasha as its first chief financial officer.

Johnson, who was both an accountant and an attorney, was a superb addition to Menasha's management team. Smart and capable, and a natural leader, he worked closely with Mowry, Sr., for the next six years, succeeding him in 1961 when brothers Mowry Smith, Sr., and Carlton Smith and their friend, Don Turner, all retired at the same time.

Many privately owned businesses come apart at the seams as they get bigger, unable to make the transition from management by a few members of the founding family to the development of a more broadly based management group. Or there may be

Dick Johnson was Menasha's president and CEO from 1961 to 1981. At left is Menasha's current corporate headquarters in Neenah, Wisconsin, completed in 1967.

knock-down-drag-out disputes within the family as to which son or daughter will succeed to the chief executive position.

Menasha has avoided these problems, and the reason dates back to Dick Johnson's presidency. His 20-year tenure marked a critical rite of passage for Menasha — from being a relatively small, family-managed business to becoming the global, professionally managed company it is today. "The family was fortunate to attract Dick Johnson," says Bruce Schnitzer, a Menasha director since 1974. "He was gracious, smart and tough, and he guarded and built the company as if he were family himself."

Johnson enjoyed the total confidence of the Smiths and Shepards. This confidence allowed him to hire seasoned executives from outside the family to fill key management positions — individuals such as Bernard "Mac" McCarragher, who joined Menasha in 1962 as budgets and accounting procedures manager and later became chairman, and William Griffith, who came

M owry Smith, Sr., left, moved up to chairman of the board in 1961 after retiring as president and chief executive officer. He chats here with his son, Mowry Smith, Jr., and with Donald C. "Tad" Shepard, Jr., and Dick Johnson.

on board in 1969 as tax and financial analyst and is today a corporate vice president and chief financial officer. Johnson and the Smith family shareholders recognized that broadening Menasha's management team was imperative if the company were to survive and grow in an increasingly competitive business environment.

"Professionalizing management was part of what we did to retain a family company," according to Tad Shepard.

CEO Bob Bero says, "I have talked with many people who are involved in privately held companies, and in my opinion we have a very rare blend today of family ownership and professional management that is totally in harmony and balance. That began with Dick Johnson. He had a huge influence on this company."

Johnson Joins Menasha

Richard Louis Johnson was born in 1916 in Madison, Wisconsin, and attended the University of Wisconsin. He received a bachelor's degree in 1939 and continued at the university to earn a law degree in 1942. For the next two years, he was an attorney with the Internal Revenue Service in Washington, D.C., returning to Wisconsin in 1944 as tax manager of Marathon Corporation, headquartered at that time in the city of Menasha. Marathon was a leading carton and paper manufacturer that later merged with American Can Company.

Because Marathon Corporation and Menasha Corporation were located near each other and were in similar businesses, Dick Johnson and Mowry Smith, Sr., got to know one another professionally and socially. By the mid-1950s, Johnson felt stuck in his career at Marathon, so when Mowry Smith offered him a job, he accepted without hesitation. "He respected Mowry an awful lot, and he thought it was a privilege to be with Menasha when it was beginning to grow and there was so much to be done," his widow, Virginia Johnson, recalls.

Johnson's first task at Menasha was to modernize its financial practices. "Until Dick Johnson came along, we were informal in our management approach," according to Al Stinchfield. "We did not employ budgeting, strategic planning or any of the other modern business practices that are taken for granted today. Dick got us started."

Johnson also spearheaded major investment programs, notably the North Bend paper mill. "Dick was a transactional genius," Griffith says. "He had the ability to structure complex projects and take them to their successful completion." His organizational talent and attention to detail, together with his background in both accounting and law, gave Johnson an unusually strong set of skills in this regard.

Forming a Management Team

Having demonstrated his management competence, Johnson was the natural choice to succeed Mowry Smith, Sr., when the latter retired in 1961. In fact, Mowry, Sr., personally selected Johnson to be the new president and CEO, bypassing fourth-generation Smiths who were then in their 30s and early 40s.

Fourth-generation Smiths did move into senior management positions. Mowry Smith, Jr., and Tad Shepard were both vice presidents and members of the board of directors when Johnson became president. Their cousin, Oliver C. Smith, was elected to the board three years later in 1964. Oliver's brother, Tamblin C. Smith, had been a director since 1954 and was elected

William Lansing joined Menasha as a research forester in 1970 after graduating from Yale with a master's degree in forest science. He became a Menasha vice president in 1983 and is today group president, Forest Products.

a vice president in 1970. (Oliver and Tam are Carlton's sons.) Johnson, however, was clearly in charge.

Non-family members of his management team included Ralph Suess, treasurer; George Hinton, general manager of the Anaheim container plant; Ernie Manders, general manager of the North Bend mill; and Nelson Page, sales manager of the Container Division. When Suess retired in 1967, Lucile Miller, who had joined the company in 1928 as a secretary, serving later as a bookkeeper and then an accountant, succeeded him as treasurer, becoming the first woman officer in Menasha Corporation's history.

In 1965, Johnson recruited Richard Widmann from Marathon Corporation to become Menasha's corporate personnel manager. With his upbeat personality and ability to get along with anybody, Widmann quickly became a popular member of the Menasha management group. He was elected vice president, corporate personnel, in 1982 and spent 26 years with the company, helping carry forward its people-focused values.

Johnson was a trim, dignified-looking man who dressed impeccably and was gentlemanly in manner. Work and family were his passions. He and his wife, Virginia, had two sons, Gregg and Timothy, to whom he was devoted. He had no hobbies until he retired, when he took up golf.

Johnson's Hands-On Management Style

Dick Johnson and Mowry Smith, Sr., could not have been more different in their management styles. Whereas Smith knew every employee by first name and made decisions based on personal knowledge and instinct, Johnson used budgeting and financial analysis as his primary management tools. He reviewed each facility's objectives and its performance against those objectives, and he held managers accountable. Even the tiniest detail did not escape him. Bill Lansing, president, Forest Products Group, recalls submitting a budget one year that included a relatively small dollar amount for the purchase of a truck. His next year's budget inadvertently repeated the item. Johnson immediately noticed the duplication and pointed it out to Lansing. "He had a grasp and a memory that were unbelievable," Lansing says. Or as Mowry Smith, Jr., puts it, "He loved detail. He could never have too much detail. But he could also see the big picture. An unusual individual."

Johnson was held in the highest regard by those who worked closely with him. His intelligence and integrity were unquestioned. To colleagues, he was a friend as well as a business associate. However, those in the company who did not work directly with Johnson sometimes found him intimidating because of his reputation as a demanding boss.

Johnson was always ready to take charge. McCarragher remembers going to lunch with him one day when Menasha Corporation was still located at its original site on the Fox River. They were riding in Johnson's car, and the Menasha president turned onto Washington Street without waiting for a break in the traffic. McCarragher relates, "A policeman who was standing on the sidewalk stopped us and yelled, 'What do you think you're doing.' Dick shot right back, 'Why the hell aren't you directing traffic? Look at this traffic, you should be taking care of it.'"

According to McCarragher, the policeman said, "Yes, sir," and immediately took up a post in the middle of the intersection. McCarragher says, "Dick got the funniest smile on his face as we drove away."

Confrontation was not, however, his usual style. "Very demanding, sometimes unreasonably so, yet very appreciative of the things you did," Al Stinchfield recalled. "And he had a fine sense of humor. You couldn't help but like him."

As was true of many business people who grew up during the Depression, Johnson was financially conservative and thrifty with the corporate dollar. "All of us at Menasha were pretty darn frugal at that time," Tad Shepard says. "None of us ever flew first class. We stayed at inexpensive hotels, and the younger people were expected to stay two to a room." Johnson also emphasized that employees who traveled to Menasha's West Coast operations should stay for several days and get a lot done, not waste the cost of air fare with one-day visits.

In addition, Johnson believed in maintaining a pristine balance sheet. Menasha was virtually debt-free during his 20-year tenure, indicating a desire to grow prudently, not aggressively. His conservative approach fit with the needs of Menasha's shareholders, many of whom were more concerned with protecting their investment than with squeezing every possible dollar of earnings and dividends out of it.

Johnson was also an outspoken advocate of American free enterprise and was troubled by the expansion of government in the 1960s and 1970s. Of course, he was not alone in that regard. In 1975, he wrote an article titled "A Lament on the State of Taxes," in which he told employees, "Congressman Vander Jagt of Michigan recently made a very interesting analysis about the high cost of government. He found that one-third of our nation's work force depends upon government-inspired pay checks, and that one-sixth work directly for the government. Perhaps the next time the 'government' wants to do something for us we should ask about that hidden kicker — what will it cost me?" He urged employees to write their elected representatives in Washington to express concern about rising taxes and increased government intervention in the private sector.

The Fire That Lit the Company's Fire

Johnson and his team took over a company that had sales approaching $30 million a year but was only marginally profitable. Its major production facilities included the Menasha and Anaheim box plants, the North Bend and Otsego paper mills, and the North Bend plywood mill. In addition, the Pacific Northwest timberlands were an important and profitable part of the business, accounting for just under 15 percent of sales. The company also owned a 51 percent interest in G. B. Lewis Company, the plastic container manufacturer that later became a wholly owned Menasha subsidiary.

DICK JOHNSON AND THE "PROFESSIONALIZATION" OF MENASHA

In 1964, Menasha's original corrugated box plant on the Fox River was destroyed by fire. Below, Tad Shepard, Dick Johnson and Nelson Page (vice president, sales) look on grimly as flames shoot into the sky. Although the fire was traumatic at the time, it benefited Menasha in the long run because the old facility was soon replaced by a new state-of-the-art plant.

AN ODYSSEY OF FIVE GENERATIONS

Menasha's corrugated box plant in Anaheim, California, served the needs of the Southern California citrus and strawberry industries, as well as of countless consumer products companies, including Oscar Mayer, Kimberly-Clark and General Mills.

One of Johnson's first priorities was to begin catching up in the corrugated container business. Menasha had grown in corrugated since World War II, but not nearly as rapidly as its larger rivals. When Johnson became president, in rapid-fire order Menasha purchased Triangle Container Company of Chicago in 1961; opened a corrugated sheet plant in Medina, Ohio, in 1962; acquired a box plant in Coloma, Michigan, in 1966; and bought Vanant Packaging Corporation (now the Sus-Rap Division) in 1968. The company also made a major investment at the Otsego corrugating medium mill, expanding its capacity with the 1965 startup of a new paper machine, nicknamed "Big Mo" number one, replacing a machine that dated to the mill's founding in 1887.

These and other investments during the 1960s were part of an ongoing effort to modernize facilities, accelerate the company's revenue growth and improve its earnings. There were many successes. However, there were failures too. The Triangle acquisition, in particular, was a mistake. The plant was unprofitable from day one of its purchase by Menasha and ended up being closed in 1965, just four years after having been acquired. Triangle was a humbling experience for the new management team.

Overhanging Menasha throughout the early 1960s was the inefficiency and dwindling profitability of its original corrugated container plant, still in operation in that onetime butter tub warehouse on the banks of the Fox River. Although the four-story factory was badly outdated, building a new plant to replace it would be costly — and Menasha was not ready to take on that expense.

As the company struggled for an answer, McCarragher remembers walking through the facility in the summer of 1964 with Johnson when the latter looked about and quipped, "What this place needs is a good fire." His offhand remark, made in jest, proved to be incredibly ironic.

The Menasha box plant's end came suddenly, dramatically and spectacularly on July 17, 1964. At about 2 p.m., a railroad employee, using an acetylene torch to repair the tracks on a wooden trestle next to the plant, inadvertently set fire to the trestle. As the flames spread to the factory, firemen arrived and brought the blaze under control. However, embers were still glowing late that afternoon when a stiff breeze arose, fanning those embers into a conflagration that engulfed the building and shot flaming sheets of corrugated into the sky. By dusk, the box plant, a mainstay of Menasha Corporation for 37 years, was mostly a shell. Fortunately, the corrugator and paper storage were in separate buildings and, therefore, not damaged.

It was a time of grief and concern for the people of Menasha Corporation. Mowry Smith, Sr., who was so well liked by employees, had died just 13 days earlier. The fact that the plant did not burn until after his death was, in a way, a blessing. He had insisted in the 1920s that the company go into the corrugated box business, and he loved the plant and admired the people who worked there. "It would have been very sad for Mowry Smith to see the plant in flames," says Walter Sellnow, a longtime employee, now retired, who was manager of the company's Wisconsin Container plant.

Many Menasha Corporation employees of that era still remember vividly where they were and what they were doing when the box plant caught fire. Of course, many were at their jobs in the plant. (Only one was seriously injured — Clarence Goeser, who spent three days in the hospital and recovered without complications.) Mike Muntner was at home, about two miles from the facility, when he heard the town fire alarm and went outside to scan the horizon for smoke. Seeing it was coming from the vicinity of the company grounds, he got in his car, turned on the radio and headed in that direction. A radio report said a Menasha Corporation warehouse was on fire, and he thought to himself, "No big problem." After all, a warehouse fire is not normally calamitous. It seldom stops production. Arriving at the site, he discovered the box plant, essential to production, was in flames. He was devastated, not only because of its potential impact on the company but also for personal reasons. Just days earlier, he had been promoted to second shift superintendent at the box plant and was looking forward to his new responsibilities. Keg Kellogg, a production manager who later became vice president of the Container Division, was playing golf when someone rushed onto the course to tell him the news. A man of few words, Kellogg walked off the course, got into his car and drove to Menasha to see what he could do to help.

Many employees were heroes. Among them was Delores Anderson, who rescued the just-completed pay checks from the burning building. Throughout the fire, Marie Kellet kept operating the switchboard by candlelight, passing on messages, receiving offers of help from other companies and re-routing calls from unknowing customers. "Everybody who could get near the place pitched right in and helped carry out records and did whatever else they could to help," Muntner says.

Members of the corporate staff pose in front of the old office building shortly before it was closed in 1967. (For a photo of the building's interior, see page 65.) Left to right are Tad Shepard, Dick Johnson, Rae Peppler, Joe Jankowski, Jan Smith, Nancy Linsmire, Lucile Miller, Phyllis Bender, Pam Zenefski, Audrey Schultz, Ray Golden, Len MacKenzie, Dorothy Britzke, John Dombeck, Mac McCarragher, Mowry Smith, Jr., Dorothy Anderson and Neal Jack.

The Move to Neenah

The fire occurred on a Friday afternoon. With the box plant in ruins, many employees wondered whether they would still have jobs on Monday morning. Those fears were quickly allayed as Menasha kept producing boxes without missing a beat. "There was never any doubt about continuing in the box business," Tad Shepard says. "We gave no thought at all to getting out of the business after the plant burned down."

Like neighbors coming to the aid of a family whose house has just been gutted by fire, other box companies in the region immediately stepped forward and volunteered to let Menasha use their manufacturing facilities during off hours to make corrugated containers. Of the 300 employees at the box plant, only 12 had to be laid off temporarily. The others were back at work on Monday in cleanup operations or making Menasha products at other companies' factories.

Though traumatic at the time, the fire was a godsend. It destroyed the company's inefficient Menasha box plant and allowed the company to replace it with a larger, highly efficient, state-of-the-art facility. What is more, the fire provided some $2 million of insurance proceeds to help pay for the new facility. As Oliver Smith points out, "The fire was dramatic, not only as an event but in the way it made a decision for us."

Literally within days, Johnson — with the able assistance of Keg Kellogg — was developing plans for a new corrugated box factory on a 30-acre tract in the neighboring city of Neenah.

Meanwhile, construction crews worked feverishly to reestablish temporary production at the burned-out plant so customer orders could be filled without interruption. Rebuilt in less than two months, and re-roofed one story high (compared to the destroyed factory's four stories), the temporary facility produced and shipped Menasha containers around the clock until the new plant in Neenah was ready.

The Neenah plant began operation in 1966 and continues today as a core Menasha Corporation facility. Muntner, the second-shift superintendent at the old plant, became superintendent of the new facility. He says, "Everything was more efficient. The layout was better and the equipment was better. We'd never seen a rotary die cutter prior to that. We'd never seen a curtain coater before. Believe me, we worked hard in that new plant to get it going. The plant was built to produce 40 million square feet of corrugated a month. In no time, we had it up to 60 million square feet. We had great people. That's what made a difference."

A year later, in 1967, Menasha Corporation completed a new corporate headquarters building at the Neenah site, replacing an antique structure on the Fox River that had housed the company's corporate office for more than 80 years. The Neenah building continues today as Menasha's corporate office.

The 1964 box plant blaze has been described as "the fire that lit the company's fire" — an event that helped modernize Menasha's manufacturing operations and, in doing so, rekindled the company's growth. In the process, it changed Menasha from an inefficient producer to an efficient producer of corrugated containers. The fire was a signal episode in the company's history.

Changes at the Timberlands

There is an old saying that problems come in bunches. That certainly was true for Menasha in the mid-1960s. Simultaneous with the aftermath of the box plant fire, Menasha's timber business (and plywood mill) in the Pacific Northwest, a major source of earnings for the prior decade, suddenly turned unprofitable because of plunging timber prices.

Unsure of how to respond, Menasha retained a professor from the University of Washington to recommend a course of action. He concluded that the timber market was unlikely to recover in the foreseeable future, and he suggested that Menasha sell all its West Coast lands and manufacturing facilities before prices went even lower. Unwilling to accept his recommendation, the company retained a management consulting firm, which came to the same conclusion. Al Stinchfield, who headed Menasha's timber operations, then phoned the dean of the School of Forestry at the University of California at Berkeley. He, too, recommended that Menasha sells its timberlands.

Despite all this advice, Menasha decided to retain the timberlands. The Smith family's emotional attachment to the timberlands was simply too great to part with them. The company did, however, trim costs by closing the North Bend plywood mill in 1967 and closing its logging operations (acquired from Irwin-Lyons) in 1969.

With the cessation of logging operations and divestment of the plywood mill, the company was back to the approach it had taken before — selling logging rights to others. However, unlike earlier in the century when the company was a passive seller responding to inquiries, this time it launched a regular schedule of competitive auctions. On any given tract, the winning bidder was responsible for cutting and removing the trees while Menasha was responsible for replanting the land. "We became a contract administrator and guardian of Menasha's timber assets," in the words of Bill Lansing, president, Forest Products Group.

This new strategy was a huge success, not so much because of the strategy itself, but due primarily to a controversial decision by the federal government. As fate would have it, just as Menasha launched its competitive auctions, the government started closing federal lands in the Pacific Northwest to logging, causing timber supplies to decline and prices to soar. The entire infrastructure of the timber industry in the region was based on the premise of a regular harvest of timber from federal lands. With access to those lands now blocked, sawmills were forced to scramble for new sources of supply. Menasha's lands, being privately owned, were not affected by the government moratorium. "All of a sudden, buyers were lined up at our door, 15 deep, wanting to bid on our timber," Lansing recalls.

Consequently, timber sales — a money-loser in the late 1960s — once again became a major source of earnings and cash flow for Menasha in the 1970s, demonstrating that prime timberland can be an excellent long-term investment so long as the owner doesn't panic when prices are down.

Board of Directors

As these major business changes were taking place, Dick Johnson focused also on the critical issue of corporate governance. Johnson was always concerned about managing Menasha Corporation with the same discipline required of a publicly owned corporation. In support of that objective, in the early 1970s he began to restructure the Menasha board of directors to give it a true oversight role.

Until that time, the Menasha board had consisted of family members and senior employees. It was a figurehead body that exercised no real power, deferring in all matters to the CEO. Seeking to strengthen the board, in 1971 Johnson recruited Menasha's first "outside" director — John Goode, a prominent local industrial consultant. In 1974, Johnson and Mowry Smith, Jr., asked Bruce Schnitzer, a young investment banker at Morgan Guaranty Trust Company in New York, to join the Menasha board as its second outside member. "I was doing merger and acquisition work at Morgan, and I think they wanted the perspective of someone who was active in the financial world," Schnitzer says. He accepted their invitation and continues on the board to this day.

Those early changes led ultimately to today's policy of having a board that is comprised approximately one-half of members of the Smith family and one-half of independent directors who are not members of the family or employees of the company. Board members have become a valuable resource for the company. An example is Dick Clarke, who joined the

The Menasha board of directors visited the company's Otsego, Michigan, paper mill in 1964. Left to right are Tad Shepard, Mowry Smith, Jr., Dick Johnson, Roman Suess (general manager of the mill), Tamblin Smith, Oliver Smith, Donald Turner, Sr., and Donald C. Shepard, Sr. A 43-year employee, Suess was one of five members of the Suess family who worked at Menasha Corporation.

Menasha board in 1984 when he was a senior executive of Celanese Corporation, heading that company's specialty chemicals and plastics businesses. Menasha was beginning to expand in plastics, and Clarke provided insight and guidance as it did so. In prior years, before it expanded its board, Menasha would never have had access to that kind of high-level expertise and advice.

Kirby Dyess, vice president of Intel Corporation, who has been a Menasha director since 1996, points out that the Menasha board does more than provide general guidance. It actively oversees management's strategies and investment plans, a far cry from the passive board of years gone by. She cites management's decision in 1996 to form ORBIS, a returnable packaging business. "Logistics was essential to the success of this business," she notes, "and the board discussed whether Menasha had the logistical capability that would be needed." The board concluded it didn't. With the board's encouragement, Menasha recruited specialists with the necessary skills.

Dick Johnson also started Menasha down the road of disclosing its financial results to employees. Until the 1950s, Menasha's results were such a closely guarded secret that even non-family senior executives didn't know how much the company earned.

Johnson changed the company's mindset from secrecy to openness. Because employees were expected to contribute in their work to the achievement of favorable corporate financial results, Johnson believed they had a right to know what those results were. That philosophy has evolved into today's policy — unusual for a private company — of sharing virtually all financial data with employees, right down to Menasha's monthly sales, earnings and return on assets.

Seeking Higher Growth

Most importantly, Dick Johnson's 20-year presidency was characterized by a relentless effort to improve Menasha's financial performance and find new ways to grow.

Acquisitions during his tenure included, among others, John Strange Paper Company and its Appleton Manufacturing and Wisconsin Container Divisions in 1969; Hartford Container Company of Hartford, Wisconsin, in 1972; Crown Corrugated Containers (a predecessor of today's Yukon Packaging Division) of Greensburg, Pennsylvania, in 1974; and Dare Pafco Products Company of Urbana, Ohio, in 1980. Newly constructed facilities included a corrugated container plant in Lakeville, Minnesota, and the Convoy pallet plant in Menasha. Remarkably, every acquisition and capital investment made during his presidency was paid for with cash, not debt.

There were also some offbeat investments that are today long forgotten — mercifully so. One involved an outfit called Anadromous, Inc., founded in 1976 by two entrepreneurs to breed salmon for sale to the food industry. Menasha invested $250,000 to obtain a 51 percent interest in the new company and leased land by its North Bend mill to the operation. According to the minutes of a Menasha board of directors meeting, "It was agreed that this was a high-risk capital investment, but perhaps worth the trouble and expense if it works out as anticipated." The entrepreneurs purchased 2.4 million salmon eggs and built a breeding pool, with a channel to the Pacific Ocean so the fish could get to open water. They figured if just two percent of the salmon returned each year to spawn, the business would be profitable. Unfortunately, only one percent returned, and the business was a bust.

Menasha also owned citrus groves in Southern California, obtained primarily from customers of the Anaheim box plant who had defaulted on their bills. The operation was called Golden Valley Groves. Menasha grew lemons, oranges and other fruit, but never made much profit. Nevertheless, there was a reason for keeping this investment. Tam Smith, a fourth-generation Smith, was administrative manager of the Anaheim box plant and headed Men-Cal Corporation, a Menasha subsidiary that collected waste paper and sold it to mills on the West Coast. He also ran Golden Valley Groves and loved that business. Keeping this relatively small operation was one of the ways in which the Smiths maintained peace in the family by ensuring that each family member active in the company had an area of personal interest.

To be fair to Dick Johnson, investments like Anadromous were peripheral. His major focus throughout the late 1960s and well into the 1970s was to consider what fundamental change, if any, should be made to maximize the value of Menasha Corporation for its shareholders. The company was still relatively small (though growing), and its earnings remained inadequate. He was searching for some bold stroke to lift Menasha to a higher level of performance, a search that eventually embraced an array of possible new directions.

One of the first ideas was that Menasha might raise additional capital, which could be used for expansion, by "going public." This idea came up initially in 1969, when the board of directors approved a resolution that Menasha consider selling up to 20 percent of its stock to public investors. Menasha executives visited Smith, Barney & Co., Goldman Sachs & Co. and other Wall Street firms to explore the concept further. Although each firm said it would be pleased to underwrite an offering of Menasha stock, the board could not reach a consensus, and the matter was tabled.

The next year, 1970, Menasha retained another Wall Street firm, Morgan Stanley & Co., to explore the possibility of a merger. Morgan suggested several potential merger partners, including Diamond Crystal Salt

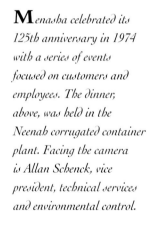

Menasha celebrated its 125th anniversary in 1974 with a series of events focused on customers and employees. The dinner, above, was held in the Neenah corrugated container plant. Facing the camera is Allan Schenck, vice president, technical services and environmental control.

Company and Associated Brewing Company. Again, the board deferred action.

Three years later, in 1973, the board of directors voted that going public or merging with another company should be "a top priority," and it formed a merger committee to research the matter, with Morgan Stanley continuing as adviser. The Menasha board looked at nearly a dozen merger and acquisition candidates in 1973 alone. Talks reached an advanced stage with Oglebay Norton Company, a Great Lakes shipping and coal company headquartered in Cleveland. However, as much as Menasha and Oglebay Norton may have wanted to merge, they could not agree on terms.

Continuing its search for a merger

partner, in 1975 Menasha purchased $1.8 million of common stock of Olinkraft Inc., a paper, packaging and wood products company listed on the New York Stock Exchange. Olinkraft was viewed as an undervalued investment as well as a merger candidate. When approached by Menasha, Olinkraft management said it had no interest in merging with anybody. Though a merger was not in the cards, Olinkraft's stock proved to be a great investment. Within less than a year, the market price of the stock had soared nearly 50 percent. Menasha then distributed its Olinkraft shares to Menasha stockholders. In this way, Menasha shareholders obtained "liquidity" — that is, the ability to raise cash by selling their newly received Olinkraft shares if they so chose. Johnson felt it was important to do this because Menasha was paying very little in the way of regular cash dividends. He believed shareholders deserved some kind of cash return, and distributing the Olinkraft stock offered an opportunity for them to obtain it.

In 1976, Menasha considered still another merger candidate, Fibreboard Corporation, a leading West Coast producer of paperboard. After intense debate, the Menasha board voted five to four not to pursue a deal. That marked the end of Menasha's seven-year, on-again-off-again search for a merger partner.

All of these proposals occurred against the backdrop of a booming economy, intensifying global competition, and a merger and acquisition frenzy that was transforming many U.S. corporations into large conglomerates. Menasha was a relatively small, private company in a business world that was changing rapidly.

Concurrent with looking at merger candidates, Menasha's directors explored the possibility of building another paper mill, a huge investment. One outside director insisted Menasha had a "linerboard void" — that is, it did not produce linerboard to meet the needs of its corrugated box plants. Linerboard had to be purchased from other companies, and this was viewed as a problem by some directors, although not by others — and certainly not by those Menasha executives involved in the company's linerboard trading program.

The board considered and rejected proposals to invest in a mill in Costa Rica and to build a mill in Texas or Oregon with a joint venture partner. The issue finally came to a head in 1976, when the board retained an engineering firm to study the economics of constructing a joint venture linerboard mill. The engineering firm concluded that Menasha was unlikely to earn much profit on such a venture, and the directors dropped any further discussion of the matter.

A Key Decision to Stay Private

This was a difficult and unsettling period for Menasha Corporation's management and directors. Although revenues advanced from $138 million in 1975 to $240 million in 1979, earnings were unchanged — and inadequate — at $11 million in both years. Return on equity declined. Moreover, despite its repeated attempts to find some way to maximize value for shareholders, the company had not completed a merger or other major deal and was back to square one.

In 1977, at the urging of Schnitzer, the Wall Street banker who had joined the board three years earlier, Dick Johnson decided the time had come to step back, look at all the options available to the company and chart a long-term course of action. The board thereupon considered five possible strategies:

1. Become a bigger factor in the paper and forest products industry by building a mill, aggressively buying more timberlands, merging with another company and/or going public to raise capital.

2. Become a specialty company in the packaging business by investing in additional corrugated box manufacturing capacity and in non-paper packaging.

3. Use Menasha's cash resources to diversify into new businesses "by acquiring control of one or more good companies," in the words of the board of directors' minutes.

4. Liquefy the company's assets "so that shareholders can realize their wealth" — that is, sell the company.

5. Continue as a private company, seeking greater growth in existing businesses and paying a higher dividend.

When the matter was framed in terms of those five basic options, the answer became clear to the board. With only a modicum of debate, it approved option number five. This critical vote meant Menasha would remain family owned, but with a more aggressive growth strategy. The months and years of soul-searching had resulted, at last, in a definitive course of action.

The Weyerhaeuser Transaction

In 1980, Dick Johnson moved up to chairman of the board while continuing as chief executive officer, and Tad Shepard, a great-grandson of Menasha founder Elisha Smith, became president and chief operating officer. Johnson spent the next year, before turning the CEO position over to Shepard, planning, implementing and completing one of the most complex and innovative transactions in Menasha Corporation history. It was exactly the type of undertaking at which he excelled.

The transaction involved the sale of Menasha's North Bend mill and Anaheim box plant to Weyerhaeuser Company for approximately $68 million of Weyerhaeuser stock. The transaction was structured so as to distribute the Weyerhaeuser stock to Menasha shareholders in a tax-free exchange — that is, the shareholders did not have to pay any tax on the Weyerhaeuser stock they received unless and until they sold it. Some members of the Smith family still own their Weyerhaeuser shares today.

Even though the sale of the West Coast mill and box plant reduced Menasha's annual revenues by 23 percent and earnings by 35 percent (totally opposite the company's desire to grow), there were valid business reasons for the transaction. "Dick was concerned that the West Coast corrugated industry was becoming more integrated and competitive," Mac McCarragher says. Other corrugated box manufacturers on the West Coast were growing in size and developing their own sources of corrugating medium and linerboard. Menasha faced the challenge of competing in an industry of West Coast giants, a challenge that would become especially difficult if these giants no longer needed to buy corrugating medium from the North Bend mill. Menasha chose to pick its battles — that is, compete aggressively in those businesses and markets where it had strengths, not butt its head against the wall attempting to compete in markets where other companies had a clear advantage.

The sale was accomplished by transferring all of Menasha's assets, except its West Coast facilities, to a new company called Menasha 1980 Corporation. The stock of Menasha 1980 Corporation was then disbursed to Menasha shareholders. As a result, shareholders held stock in two companies: the original Menasha Corporation, which now owned only the North Bend mill, Anaheim box plant and related West Coast facilities, and Menasha 1980 Corporation, which owned all the rest of the company's assets.

Menasha Corporation was then acquired by Weyerhaeuser. Having been acquired, Menasha Corporation no longer existed, and Menasha 1980 Corporation changed its name to Menasha Corporation.

Thus, to be legally precise, today's Menasha Corporation was established in 1980 and was once called Menasha 1980 Corporation.

In addition to the competitive reasons for the transaction, the Weyerhaeuser deal had a purpose similar to the earlier Olinkraft transaction: it provided Menasha shareholders with liquidity. One of the thorniest problems faced by family companies is the estate taxes that must be paid as ownership of a company passes from one generation to the next. Frequently, companies are forced to go public, whether they want to or not, so that family members can raise cash by selling stock to pay these taxes. For Menasha shareholders, the Weyerhaeuser transaction was an alternative to taking Menasha public in that it gave them Weyerhaeuser stock which they could sell if and when they needed cash for estate taxes or any other purpose. "The Weyerhaeuser transaction made it possible to keep Menasha private," Schnitzer observes.

During his two decades at the helm, Dick Johnson had brought Menasha Corporation into the modern era by developing rigorous financial controls, hiring talented managers from outside the family, divesting money-losing operations such as the North Bend plywood mill, investing in new businesses such as plastics and structuring an unusual transaction that fostered Menasha's continued private ownership. As the decade of the 1980s began, it was up to Tad Shepard to build on those accomplishments and launch the company on an aggressive growth path.

Menasha's First Punch-Card Computer

In 1965, Menasha took its first tentative step into the modern computer era. In fact, most companies were just beginning to switch from manually operated calculators to automated calculating systems.

The company hired Elder "Al" Firgens from Northern Paper Mills to supervise the newly formed IBM Department. "The function of the department," a company announcement said, "will be to process source data information and present facts in a printed format."

Menasha outfitted Firgens and his staff of two key punch operators with an array of equipment rented from International Business Machines Corporation, including two key punch machines, a card sorter, card reader and automated calculator. "It is extremely difficult in this industry to determine costs (particularly for small orders) and, consequently, to determine if an order would bring a profit," the announcement stated. "With the use of IBM punch cards, each item of reported production can be used to determine production costs for each order and compared to selling prices. The same information can be used to compute incentive payrolls and production statistics."

This punch-card system was incredibly primitive compared to today's workplace environment in which powerful PCs and other forms of automation are so omnipresent they are taken for granted. Nonetheless, it was state of the art for its time.

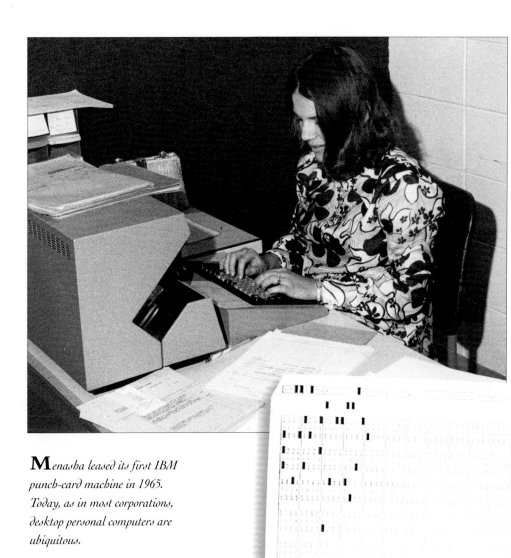

Menasha leased its first IBM punch-card machine in 1965. Today, as in most corporations, desktop personal computers are ubiquitous.

CHAPTER EIGHT

The Fourth Generation

Tad Shepard took charge of a company that was suddenly one-quarter smaller in revenues because of the Weyerhaeuser transaction, yet had ambitious growth plans. And grow it did.

In 1982, Shepard's first full year as CEO, Menasha's sales were $195 million and earnings were $7 million. Seven years later, when he moved up to chairman, handing over the CEO title to Walter H. Drew, sales had reached $545 million and earnings were $40 million, one of the greatest growth spurts in Menasha Corporation's history.

During that same period, return on equity advanced from 7 percent to 19 percent, and the company went from 2,000 employees and 31 plants in 1982 to 3,900 employees and 38 plants in 1989.

These results reflected, to some degree, the benefits of an improving U.S. economy. Even more important, however, was the aggressive growth strategy implemented by Shepard and his management team. Speaking of Shepard's management style, Bill Griffith says, "Tad is very smart, very focused, very results-oriented, doesn't believe in a lot of bureaucracy, doesn't waste time, gets the job done."

Donald C. "Tad" Shepard, Jr., a great-grandson of Elisha Smith, was Menasha's president and CEO in the 1980s. Opposite, cousins Mowry Smith, Jr., left, and Clark Smith visit a "trade show" of Menasha products at the company's annual shareholder meeting, held each spring.

How He Joined the Family Company

Donald Carrington Shepard, Jr., a great-grandson of Elisha Smith and grandson of Charles Smith, was born in 1924 and grew up in Neenah. His younger brother, Charles Shepard, who became a schoolteacher and later a headmaster, recalls their childhood as being idyllic and "brimful of Smiths." Their uncles, Mowry Smith, Sr., and Carlton Smith, lived nearby. The three third-generation Smith siblings — Mowry, Sr., Carlton and Tad's mother, Sylvia Smith Shepard — and their families saw each other regularly and sometimes gathered for holiday dinners. The Shepard household was a beehive of activity, with friends, relatives and business associates dropping by constantly.

Although his father was a Menasha Corporation vice president and director and his parents talked often about the family business, it was not a foregone conclusion that Tad would work for the company when

Tad Shepard, left, dines with Bob Briggs, a Menasha executive in charge of corrugated container marketing and sales, and with Tad's father, Don Shepard, a Menasha vice president, at the Fibre Box Association meeting in Boca Raton, Florida, in 1956. Tad Shepard served later as the association's president.

he grew up. His parents, he says, neither encouraged nor discouraged him to do so, believing he should make up his own mind.

Shepard attended prep school at Hotchkiss in Connecticut and, on graduating in 1943, enlisted in the Marines, serving in the Pacific as a navigator on a B-25 bomber. After the war, he enrolled at Yale. While an undergraduate, he met Jane Steinborg, a Barnard College student, at a Yale-Princeton football game. They were introduced by Shepard's cousin, John Sensenbrenner, Jr., a grandson of Henry Smith. Jane says she was "nearly engaged" to another man at the time. But Shepard was drawn to her immediately, courted her and won her hand. They were married in 1949, Shepard's senior year. They have four children (including son Bill, who is president of Menasha's Packaging Group) and nine grandchildren. Shepard's family is very important to him. Although he does not wear his feelings on his sleeve, he does not hide them either. Not long ago, Jane said, "Tad was talking on the telephone, and I heard him tell someone he fell in love with me at first sight." A smile came over her face as she made that comment.

On graduating from Yale in 1950, Shepard interviewed for jobs at several major Midwestern companies, including 3M and Pillsbury, before finally deciding on Menasha. "The more I looked into it," he explains, "the more I thought this was an interesting opportunity."

Shepard began as a sales trainee in Menasha Corporation's corrugated box business, although it quickly became apparent there was nothing for him to sell. Federal production controls were then in effect because of the Korean War, and Menasha could not produce enough boxes to market to new customers. Shepard therefore moved almost immediately into the production side of the box business. He became a Menasha vice president in 1953, a member of the board of directors the following year, production manager of the Neenah Container Division in 1955 and general manager of Neenah Container in 1959, assuming responsibility for all of Menasha's Midwest container plants the following year. By 1968, all container plants reported to him, and in 1977 the plastics group came under his supervision as well.

From the 1950s through the 1970s, corrugated boxes were Menasha's largest business by far. "Apart from the timberlands, boxes were just about all we had," Shepard recalls. In the early part of his career, Shepard worked closely with Bob Briggs, a marketing executive who had been hired by Mowry Smith, Sr. Shepard was in charge of box production while Briggs was in charge of marketing and sales. Briggs eventually left Menasha when it became clear he was not in line for the presidency.

Four of Shepard's cousins — Mowry Smith, Jr., Tam Smith, Lawton Smith and Oliver Smith — were also active in the business. The careers of Tad and Mowry, Jr., in particular, moved on similar tracks. After receiving a bachelor's degree from Yale University and an MBA from Stanford University, Mowry, Jr., joined Menasha in 1952 at the Rockford, Illinois, furniture plant. In May of that year, he transferred to the North Bend plywood plant and four years later, in 1956, to the Anaheim container plant. In 1960, he moved to the corporate office in Menasha and was elected president of Menasha Wooden Ware Company (the investment company which owned a portfolio of stocks and bonds then worth about $15 million) and a vice

Mowry Smith, Jr., joined Menasha in 1952 and spent more than 30 years with the company, becoming a key member of the senior management team. In 1960, he was elected president of Menasha Wooden Ware Company (the investment company which owned a portfolio of stocks and bonds) and a vice president of Menasha Corporation, later serving as a Menasha Corporation senior vice president. He has a deep interest in the history of the family and the company and in 1974 wrote a book about the company.

president of Menasha Corporation. He became senior vice president in 1970, with responsibility for the Land & Timber and Plywood & Lumber Divisions, as well as for wood flour operations. Mowry retired in 1983, Tad in 1989.

Tad and Mowry, Jr., the two most influential fourth-generation Smiths in the management of the business, did not always see eye to eye. Their personalities and management styles often clashed. Many longtime Menasha employees still recall the tensions that existed between them over matters of corporate policy and business strategy. Yet, Tad and Mowry kept their disagreements from erupting into personal animosity. They remained friendly and sometimes got together socially, such as to play tennis, as they continue to do today. Part of the Smith family heritage is that family members can disagree, but must find ways to accommodate those disagreements — and never let them destroy family bonds or harm the company. Mowry, Jr., says, "It's true that we often disagreed. But we tried to keep our differences in perspective."

Overcoming the Skeptics

As Shepard advanced in his career, he and Dick Johnson began to work closely together, and it soon became apparent to others in senior management that Johnson was grooming Shepard to become Menasha's next CEO. "Dick and I developed an awfully good relationship," Shepard says. "We became friends and worked well together. However, Dick never told me I was being groomed. It just happened."

In May 1980, Shepard was named president and chief operating officer of Menasha Corporation. He became president and chief executive officer in 1981, continuing in that position until 1989, when he retired and was succeeded by Wally Drew, recruited from Kimberly-Clark.

Shepard was, in fact, a controversial choice for CEO. When his appointment was announced in 1980, many in the company questioned whether he had the experience to lead the company. Mac McCarragher, for one, says, "I have told Tad this. I thought he was going to be a disaster. As vice president of the container operations, Tad was so focused on manufacturing that I thought he would not have the perspective to lead the company. I was surprised it was the other way around. He was a very thoughtful and capable CEO."

Don Riviere, a Menasha vice president who died tragically in a 1997 plane crash, recalled in a 1996 interview that Shepard initially lacked some of the skills associated with a CEO, such as the ability to speak well before an audience. "Tad would get up to the podium, grip both sides tightly and stare down at what he was reading without ever looking at the audience," according to Riviere. He added, "In time, Tad became a much better speaker. He really broadened all his skills and was an excellent CEO."

Even Tad's brother, Charles, who

Tad Shepard and President George Bush were classmates at Yale. As is true with most college classmates, the friendship and warm feelings remain.

viewed the company from afar as a stockholder rather than an employee, says, "I have very strong feelings of pride and amazement at Tad's work, his competence, his growth in the job, his courage in taking on the job in the face of people who didn't approve of what he was doing and the success that came from that."

As for Tad Shepard himself, he says questions as to whether he was the right man for the job were no big deal to him. "I told them," he recalls, "'You're going to have to decide whether you want me. I am what I am — that's what you're getting.'" He says he had absolute confidence in his own abilities. "We faced a big challenge in trying to grow this company," he says, "and I knew just what I thought had to be done."

Shepard took charge with a flourish. Known throughout his career as an extremely hard worker, he set a breakneck pace on becoming CEO, arriving at his office as early as 4 a.m. He often spent the pre-dawn hours writing a flurry of notes referred to by the Menasha executives who received them as "T-grams." Typically, these notes asked pointed questions about recent business performance or made suggestions as to actions that might be taken to correct problems or capitalize on opportunities.

Shepard's devotion to work was so great he never took a sick day for 35 years. This unusual streak finally ended in 1986 when he was stricken with a rare virus called Guillain-Barré which paralyzed his legs and weakened much of the rest of his body. Struggling with the illness, but refusing to give in to it, in a message to employees several weeks later, he wrote, "Fortunately I have been able to use my hands quite well and continue to handle my business mail as usual." He returned to the office part time in a wheelchair seven weeks after having fallen ill. He eventually recovered fully.

Bill Lansing says of Shepard, "There is not a dishonest, unethical bone in his body." When Lansing was transferred from Forest Products to the corporate office for a year in the 1970s to broaden his experience, he says Shepard, who then headed the Container Division, told him, "I don't think you should be here, but now that you are I'm going to work your behind off." Lansing adds, "He did, and we quickly became friends."

Restoring Employee Morale

Shepard faced a number of pressing issues on becoming CEO in 1981. Foremost was the deterioration of employee morale following the sale of the Anaheim box plant and North Bend paper mill to Weyerhaeuser. The Weyerhaeuser transaction was a one-time deal, not the beginning of the sale of other Menasha facilities. Nonetheless, many Menasha employees wondered whether their operation might go next.

In a November 1981 message to Menasha employees, Shepard acknowledged the loss of many "fine employees" through the Weyerhaeuser transaction. "As Dick Johnson wrote a year ago," he

Donald Riviere joined Menasha Corporation in 1970 from DuPont Company. He later became Menasha's vice president-corporate strategic development, playing a major role in developing a more aggressive growth strategy. He and another Menasha senior executive, John Snyder, died tragically in a 1997 plane crash.

pointed out, "the reasons for the transaction are many and complicated but we are convinced it will make Menasha Corporation a sounder company. At any rate, the long and difficult period of uncertainty is behind us and we are well into the planning to build Menasha back to the sales and profit levels that previously existed." He kept plugging away at the topic, reassuring employees that Menasha intended to grow, not scale back.

Partly because of morale problems, but also because of his determination to establish a clear corporate direction, in 1982 Shepard assembled the senior management team for a weekend at a country cottage to write a mission statement. Menasha's mission, which continues to this day, is "to provide its stockholders with an investment which will provide steady growth, reasonable and dependable dividends and a return on equity better than average for all U.S. companies; to recognize a special obligation to its employees as well as its responsibility to its customers, suppliers and the communities in which it operates."

The mission statement also established excellence goals and an excellence process. The excellence goals, for instance, are:

1. We understand our business
2. We know what is expected and how we are doing
3. We have the skills to perform
4. We are involved in business progress and change

In addition to addressing the morale issue, the mission statement and excellence

THE FOURTH GENERATION

◤◣ MENASHA CORPORATION

In the early 1980s, Tad Shepard brought together the senior management team to create a mission statement and operating principles that still guide the company today.

Corporate Mission

To provide its stockholders with an investment which will provide steady growth, reasonable and dependable dividends and a return on equity better than average for all U.S. companies; to recognize a special obligation to its employees as well as its responsibility to its customers, suppliers and the communities in which it operates.

Operating Principles

To constantly strive for excellence through the aggressive development of human resources, improvement of all employees' management and job skills, delivery of superior quality and service and prudent investment of financial resources; to place top priority on honesty, integrity and fairness while preserving a work climate that rewards individual expression, ingenuity and courage.

process had a broader purpose. Don Riviere recalled, "Tad looked at Menasha Corporation and said, 'It's a good sound company, but it's not performing as well as it should.' We were in the paperboard, packaging, plastics and forest products industries, and we said, 'Gosh, isn't it awful we're in such bad industries?' Then we realized that wasn't the problem. We had to come to grips with why we weren't performing at the level we should."

One reason for this underperformance, according to Riviere, was that Menasha Corporation was not fully utilizing the ideas and talents of its employees. "So we began a major effort to do something about that, and out of this effort came the excellence process," he noted. Or, as Shepard himself puts it, the goal was to "strive for excellence through the aggressive development of human resources."

In 1984, explaining the excellence process, Menasha's employee magazine noted, "One person, by effort and will, can transform another person. We each have the ability to make every day, every moment, worthwhile so we can smile at the end of the day."

Listening to Shareholders

Another of Shepard's early priorities was to establish a stronger bond with the Smith family shareholders. Historically, the business had been run with little or no input from family members who did not work for the company. This arrangement was satisfactory in the era of Charles Smith and Mowry Smith, Sr., when so many of Elisha Smith's descendants lived in the Menasha/Neenah region and were active in the business.

By the early 1980s, relatively few of Elisha's descendants worked at the company and, in fact, a great many had left Wisconsin to pursue careers in business, law, education and other fields. For several years, Mowry Smith, Jr., had been informally eliciting the views of shareholders and presenting those views to the board of directors. However, Mowry, Jr., as well as Dick Johnson, Tad Shepard and Mac McCarragher, all felt the time had come to learn more systematically about shareholder expectations before Menasha Corporation embarked on an aggressive growth strategy.

Shepard and McCarragher took up this challenge, traveling throughout the United States to meet with small groups of shareholders. Those meetings were a turning point in Menasha Corporation's relationship with its shareholders — from virtually no involvement by non-employee shareholders to today's more active role, including a large and vocal turnout each spring for the company's annual meeting. Almost certainly, Menasha would no longer be family owned today if shareholders had not become involved.

At those meetings in the 1980s, McCarragher says shareholders expressed pride in their Smith family heritage and wanted to keep the company in family hands. Moreover, they voiced a decided feeling that Menasha continue to conduct its business in an ethical manner. McCarragher reports, "They said things like, 'We want to make sure Menasha is an equal opportunity employer,' or 'We want to be certain it is environmentally responsible.' Those kinds of reactions came back very clearly."

As to Menasha's financial performance, shareholders had few concerns or complaints. McCarragher says, "We asked, 'Would you like to see the dividend rate increased?' The basic reaction was, 'If it doesn't cause detriment to the company, sure we'd like to see it increased.' Some of the younger generation, in particular, liked the idea of a larger dividend, but it was not an urgent demand." Out of those meetings came a decision to increase the dividend to 35 percent of Menasha's earnings from 20 percent.

Some shareholders also said they would like Menasha to become more diversified. For many shareholders, Menasha was their dominant investment, and they worried about having the lion's share of their assets tied up in a company that was vulnerable to cyclical swings in the corrugated box and timber businesses.

Growth and Diversification

Buttressed by shareholder support for diversification, and working closely with Riviere and McCarragher, Shepard began to acquire high-potential firms that could improve Menasha's return on investment.

Finding the right deals at reasonable prices was time-consuming. In one three-year period during the early 1980s, Menasha reviewed more than 100 acquisition candidates, studied 70 of them in depth, negotiated with about 30, and ended up acquiring two and forming joint ventures with two others.

All told, Menasha acquired an average of about two companies per year during Shepard's presidency, double the rate of the prior two decades. Importantly, these acquisitions expanded the company's participation in plastics and took it into a new business, commercial printing.

Key acquisitions included Scranton Plastics Laminating Corporation in 1981; Vinland Corporation (web-printed paper and plastic film products) in 1982; Traex Corporation (reusable plastic products for the food service industry) in 1984; Mid America Tag & Label (product identification and merchandising tags and labels) in 1985; The Murfin Company (web screen printer of label and identity products) in 1986; Neenah Printing and Oshkosh Printers, Inc., (commercial printing) in 1986; Thermotech (precision injection molding of thermoplastics and engineered resins) in 1988; and Colonial Container Company (corrugated cartons) in 1989.

Additionally, Shepard used debt to buy companies, a dramatic change for Menasha, which had been virtually debt-free throughout the first century and a quarter of its existence. "Our first significant use of debt occurred in 1988, when we borrowed

The purchase of Thermotech in 1988 was Menasha's largest acquisition to that time. The unit is a custom injection molder of precision products, serving companies in such industries as automotive, electronics, medical and water purification.

THE FOURTH GENERATION

Left, "Big Mo" is one of two papermaking machines at the company's mill in Otsego, Michigan. Right, Menasha entered the transportation business in 1982, converting its fleet of trucks to a common carrier.

approximately $30 million to buy Thermotech," Mac McCarragher relates. Since then, Menasha has routinely used debt to finance investments.

Investing at Otsego

Shepard also invested in existing operations, including a series of major projects that literally transformed the nearly century-old Otsego paper mill in Michigan.

In 1982, Menasha installed new repulping equipment, a new cleaning system for wastewater and new chip-handling equipment at Otsego. Three years later, the mill's "Big Mo" number one paper machine was rebuilt at a cost of $5.3 million, nearly doubling its speed from 1,200 feet per minute to 2,250 feet per minute, in this way nearly doubling its capacity while sharply reducing its average unit production cost.

An even bigger investment at Otsego was launched in the late 1980s, near the end of Shepard's presidency. Costing $55 million, by far the largest capital investment in Menasha Corporation history to that time, the project increased the machine's speed a further 44 percent to 3,250 feet per minute and installed the equipment to make a high-quality 100 percent recycled product.

Allan Schenck, a retired Menasha vice president, notes that the investments at Otsego demonstrate a vital trend in the American paper industry. "The big change over the years," he says, "has been building bigger, faster machines to keep costs down. I take a lot of pride in the fact that the paper industry is one of the few major industries that has not suffered great losses to foreign competition. There is a reason for that. Competition to cut costs has been so fierce in the United States that the foreign mills haven't been able to keep up."

With these and other investments, the Otsego mill has been so extensively upgraded in recent years that its turn-of-the-century employees, if they were alive today, would not recognize the place. "The facility is, without doubt, one of the most modern and efficient corrugating medium mills in the nation," says Bruce Buchanan, president, Paperboard & Services Group.

Entering the Transportation Business

Looking for entirely new opportunities, in 1982 Menasha converted its fleet of trucks to an interstate common carrier, Menasha Transport, Inc.

Carl Kraus, who headed Menasha's transportation operations and was instrumental in starting the new business, says Menasha's trucks, after delivering products to customers, had traditionally returned empty. "We saw an opportunity to reduce the amount of empty time by transporting products for others, and to make a profit doing so," he says.

To provide greater visibility for Menasha Corporation, all the trucks in the Menasha fleet were painted green and gold on white, reminiscent of the "Menasha green" boxcars of old. Menasha Transport continues in the common carrier business to this day.

Developing an Export Market For Timber

Meanwhile, results of the timber business on the West Coast — a mainstay of Menasha's earnings for most of the 1970s, but always subject to cyclical market forces — plunged suddenly in 1979. This happened during a period of economic turmoil for the nation. As inflation increased, home mortgage rates surged to nearly 20 percent and construction of new homes came virtually to a halt, causing timber demand and prices to slump badly.

"It was like a tornado sweeping through the forest," Bill Lansing says. "We didn't see it coming. We put timber up for sale and suddenly nobody wanted to buy it. We woke up to the fact that we were at the end of a golden era for Menasha's timberlands and we had to find a new approach."

Out of that traumatic event came a decision to export Menasha timber to Asia, where demand remained strong. Establishing a market presence in Asia proved to be difficult and challenging. After much effort, it finally paid off in 1982 when the company sent its first shipload of logs — 40-foot logs worth $1.5 million — to the People's Republic of China. "We hit our stride in 1983 and have been exporting to Asia ever since," Lansing says. The company now exports primarily to Japan, where Douglas fir is prized for home construction.

"There are some very big players in the timber export business," Lansing notes, "and we had to carve out a niche to be successful." Menasha specializes in what Lansing calls the "supermarket method of marketing." Like various brands of breakfast cereal, all logs are not created equal. The company's vast log yard on the shore of Coos Bay is stacked with logs sorted by species, age, size, knots and other characteristics that determine value. "The customer can say, 'I want 50 of those logs, 50 of this other kind and 100 of that type over there,'" according to Lansing. "We put together the package of logs the customer wants, and if the customer wants a log that's not a traditional length, we'll cut it for him. We learned not to say no to our customers."

Also, in 1981, at Tad Shepard's insistence, the Land & Timber Division developed a business model for maximizing the long-term value of the company's timberlands. A traditional objective of many timber companies is to cut the same number of trees each year to sustain a given level of volume. Lansing recalls, "Tad said, 'We don't want sustained yield. We want optimum value.'" With the help of McCarragher and Riviere, the Land & Timber Division created a business model that involves harvesting more trees when prices are strong and fewer when prices are weak. The model also makes clear that younger trees cannot be cut. Such trees must be preserved and grown to ensure the high value of the timberlands for future generations of Menasha Corporation shareholders. Menasha's timber business continues to be guided by the export program and harvesting model put in place during Shepard's presidency.

Decentralized Operations

When Shepard became president, some skeptics questioned whether he had the personality and mind-set to share management responsibility with others. Board member Evan Galbraith says, "Tad moved ahead early in his career as a terrific hands-on manager. As a result, there was considerable question within the family and on the board as to whether he could delegate sufficiently to be a good CEO. But having Dick Johnson around to help Tad mature into that job, and then the way Tad blossomed incredibly, proved that these doubts were wrong."

In fact, one of the enduring legacies of

Logs from Menasha's West Coast timberlands are loaded onto a freighter in Coos Bay, Oregon, for shipment to customers in Asia. Coos Bay is one of the great natural harbors in the Pacific Northwest.

Shepard's presidency is the decentralization of Menasha Corporation management. Throughout its first century, Menasha had maintained a top-down organizational structure with authority concentrated in the hands of the CEO, who made all important decisions. Dick Johnson took the first steps toward giving general managers greater authority to run their facilities, subject to accountability for results. Shepard pushed this policy further. He believed that controlling the company from the top was no longer practical in light of growing sales volumes and the increased diversity of operations.

Shepard's philosophy continues today, with Menasha being organized into seven autonomous business groups. "These units," a company document explains, "operate within a decentralized structure that combines the advantages of local, entrepreneurial leadership with the strength of specialized corporate support."

Wally Drew

Whereas Shepard was groomed to succeed Dick Johnson, there was no clear successor as Shepard approached retirement. Thus, at a board of directors meeting in September 1986, he reaffirmed his decision to retire in 1989 and suggested the company begin planning for a new CEO.

For the next two years, the board considered various candidates, ultimately recruiting Walter Drew from Kimberly-Clark Corporation. Drew was an executive vice president of that company, heading its

U.S. consumer products business, which had sales of about $2.5 billion a year, roughly five times the size of Menasha at that time. He lived and worked in the Neenah area and was well liked and respected by the many Menasha executives who knew him. In an unusual twist, his mother, Marion Drew, had worked part time for Menasha Corporation from 1962 to 1965 as editor of the *Log*, the company's employee publication.

With his strong background in marketing at a major corporation, Wally Drew seemed like an ideal choice to succeed Shepard. "In the distant past, Menasha Wooden Ware was driven by its ability to manufacture products," according to Riviere. "Then under Dick Johnson and even more strongly under Tad Shepard, the corporation became division driven. The divisions focused on making products and the company's results were based on how well the divisions did. When Wally Drew came in, we said we needed to be more market driven and less product and process driven."

Drew says he had no hesitation about joining Menasha Corporation, even though it was much smaller than Kimberly-Clark, because he admired Menasha and its people and welcomed the opportunity to become CEO. "I figured I'd spend the rest of my career having a lot of fun running my own show at Menasha Corporation," he says. He joined Menasha in June 1988 at age 53 as senior vice president and chief operating officer, spending the next year familiarizing himself with company operations.

Walter Drew became Menasha's 10th president in 1989, leading the company for the next three years. He came to Menasha from Kimberly-Clark Corporation, where he was executive vice president in charge of its U.S. consumer products business. Menasha's goal in recruiting him was to sharpen its marketing focus.

It was with great hope and enthusiasm that Drew became Menasha Corporation's 10th president in June 1989, succeeding Shepard. Almost immediately, however, there was a clash of cultures. Drew came from a culture at Kimberly-Clark where maximizing shareholder value was the number-one priority, and to that end, he placed heavy emphasis on cost cutting at Menasha. For instance, he developed a companywide program of centralizing purchasing that went against Menasha's culture of decentralized operations. In addition, he terminated several executives who, in his opinion, were not performing well. "I was trying to change the culture," he says. "I plead guilty." He also says, "The Shepards and the Smiths view this more as a private trust that needs to be nurtured and developed and handed off to the next generation. Maximizing shareholder value isn't necessarily the number-one priority."

Drew had supporters within Menasha who believed he was right in trying to shake up the company. Others, however, bridled at his approach. By 1992, it was clear that Drew and Menasha Corporation were not a perfect fit after all. Drew recalls, "I had no trouble when Tad came to me and said, 'This isn't working.' I said, 'You're right.'"

Drew and Menasha agreed to part ways in as friendly a manner as possible. In his usual forthright manner, Shepard said in a press interview, "Wally is a gentleman and a damned good friend of mine, but he has a different management style and Menasha has a strong culture, and the two didn't fit together like a marriage made in heaven."

THE FOURTH GENERATION

Tad Shepard's family attended his retirement party in Appleton, Wisconsin. Pictured around the table, clockwise from left, are Jane Shepard (Tad's wife), Timothy Shepard (Tad's nephew and a fifth-generation descendant), Julia Shepard Waite (Tad's daughter), Donald C. Shepard III (Tad's son), Bill Shepard (Tad's son) and S. Sylvia Shepard (Tad's daughter).

In that same article, Drew was quoted as saying that he and Shepard remained friends. "There's no palace revolt going on over there," Drew assured, "there's no problem."

Following Drew's resignation, Shepard returned as acting CEO and McCarragher, who had retired the previous year, returned as vice chairman. In a message to employees, Shepard emphasized, "Wally Drew's resignation definitely does not create a crisis for Menasha Corporation." He pointed out, "Menasha is decentralized and we govern through nine vice presidents who are capable, experienced business executives. They, in turn, manage through 49 general managers and department heads. These men and women manage their businesses needing minimum guidance from the vice president heading their business group."

Speaking of Drew's resignation, Shepard wrote, "Mutually we found that his management style did not effectively mesh with Menasha's deeply ingrained culture and the best course seemed to each go our own way. Wally is a fine person and a good businessman. His three years as president brought important, beneficial changes that hopefully will be with us for many years."

Shepard served as acting CEO for a year, "re-retiring" in June 1993 when Bob Bero became Menasha's new president and chief executive officer.

The Tad Shepard Years: A Summary

Looking back on Tad Shepard's presidency, Galbraith says, "Tad brought vision to the company. The formalization of a Mission Statement, the formalization of strategic planning, the formalization of management development and management succession — all those things had been around for many years, but had never been articulated as fully as they were under Tad.

"Tad Shepard's stewardship will also be remembered as a remarkable period of growth for the company and a remarkable period of reaffirmation of the fundamental values that make Menasha Corporation special and different."

Shepard remained on the Menasha board of directors as chairman until 1995, when he retired from the board, ending a 45-year career with the company. He and his wife continue to live in Neenah and enjoy travel, tennis and other leisure activities.

CHAPTER NINE

PLASTICS

Over the course of its history, Menasha Corporation has made products from three main materials. Each is a vital part of the company's story. Moreover, taken as a group, the three materials highlight Menasha's ability to change — indeed, the necessity for change — as customer demands and material technologies have evolved.

First came *wood* — initially through the company's dominance of the woodenware business in the late nineteenth and early twentieth centuries and continuing today with the sale of logs from the West Coast timberlands. Next came *paper* — beginning with the company's entry into the corrugated container business in 1927 and continuing today not only in that business but also with the manufacture of corrugating medium at the Otsego mill. Finally came *plastics* — beginning in 1955 with a key investment in the G. B. Lewis Company of Watertown, Wisconsin, and continuing today with the Lewis operations and a host of other plastics-related businesses acquired

Menasha's array of plastic products includes the returnable, reusable packaging of its ORBIS Division, opposite, and the ultra-high molecular weight (UHMW) polyethylene used to make the cutting board above.

and developed since the mid-1950s.

Bob Bero says, "We continue to expand in plastics, not to the exclusion of our other businesses, but to capitalize on the growth opportunities we see there." In fact, three of Menasha's seven operating groups — Poly Hi Solidur, Polymer Technologies and Material Handling — are involved in the plastics industry. The plastics made by Menasha are used in end-market applications as diverse as hip replacements, returnable packaging and the fan blades inside Black & Decker "Dustbusters."

In the 1967 movie, *The Graduate*, Dustin Hoffman was advised that plastics were the future. Menasha's experience shows that plastics, with their versatility, durability and attractive appearance, are indeed a growth industry.

The Acquisition of G. B. Lewis Company

Ironically, Menasha got into the plastics business by acquiring an old-line woodenware company that, like Menasha itself, was struggling to change. That company was G. B. Lewis of Watertown, Wisconsin.

AN ODYSSEY OF FIVE GENERATIONS

Menasha got a toehold in plastics in 1955 when it bought an interest in G. B. Lewis Company, an old-line woodenware company that was converting to plastic products. The ad below extolled a Lewis wooden container that was used by American troops during World War I. The box originally belonged to the Red Cross, which filled it with sandwiches for troops training in Pennsylvania. The soldiers took the box with them to Europe and used it to transport bread. After the war, they gave the box back to the Red Cross with the words, "We return this box as a souvenir, as it has been on every battle front in France and also on the Rhine, and we thank you very kindly for its contents, as the sandwiches were fine."

Lewis was founded in 1863 by G. B. Lewis and his brother, R. E. Lewis, natives of Saratoga County in upstate New York. Journeying to Wisconsin to become farmers (not long after Elisha Smith had made a similar journey to enter the dry-goods business), they soon opened a woodenware factory on the banks of the Rock River (as Elisha Smith did on the banks of the Fox River about 70 miles to the north).

Menasha Corporation and G. B. Lewis Company were among the many woodenware companies that took root across the state of Wisconsin in the mid-nineteenth

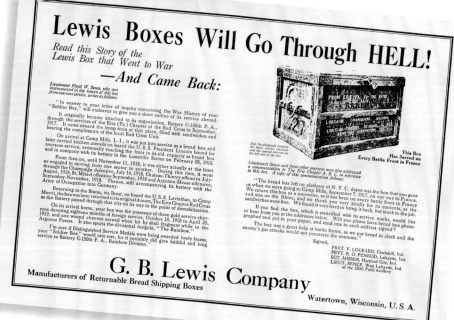

124

century. While Menasha developed into the largest of these companies, Lewis was successful in its own right, specializing in the manufacture of cheese boxes, honey boxes and beehives. The company grew under family management, then hit upon the hard times that afflicted all woodenware manufacturers in the early years of the twentieth century. Responding to the market's sudden downturn, Lewis diversified by introducing Arkitoy (a wooden construction toy similar to Erector sets) in the 1920s. In the 1930s, as Menasha was diversifying into corrugated, Lewis began manufacturing wooden slats for venetian blinds, becoming the nation's number-one producer. During World War II, the company produced laminated wooden propellers for military and commercial aircraft, continuing also to make beekeeping supplies and other woodenware products.

In 1948, Gordon Frater — a University of Wisconsin engineering graduate who had been with Lewis since 1920, having previously worked at General Electric — became Lewis's president. He was a go-getter who, in 1949, took the company into plastics, initially with the production of fiberglass-reinforced polyester containers. These containers were used by industrial companies to transport parts in-plant and from warehouses to production lines and by bakeries to deliver bread to retail stores.

Explaining why plastics were replacing wood in these and other applications, Milton Frater, Gordon's son, notes, "New plastics with excellent performance charac-

Convoy reusable pallets are one of the many products of Menasha's ORBIS Division. The division was created in 1996 by combining LEWISystems, the Convoy Plastic Pallets Division, and the DuraPAK and Donray businesses to provide complete returnable packaging solutions for business.

teristics were coming onto the market. Woodenware, which was labor-intensive and therefore costly to make, could not compete with these plastics, which could be molded or extruded into finished products relatively inexpensively."

Having launched the company in a new direction, Gordon Frater sought to buy the company from the descendants of the founders, and he turned to his longtime friend, Mowry Smith, Sr., of Menasha Corporation, for financial support. As a result, in 1955, the Smith and Frater families became business partners: Menasha Corporation acquired 51 percent of G. B. Lewis, and the Frater family acquired 49 percent. Mowry Smith, Jr., explains his father's interest in investing in G. B. Lewis as follows: "My father said, 'Well, I can see that plastics are becoming a threat to paper packaging. We ought to have a foot in the door.'"

At that time, Lewis's revenues were about $1.5 million a year, and plastics were a small but rapidly increasing portion of its sales. Lewis became the platform for Menasha's early growth in plastics.

"Menasha was interested in plastics, and they encouraged us to expand in that business," says Boyd Flater, who was G. B. Lewis Company's controller and later its president. Lewis subsequently sold its

AN ODYSSEY OF FIVE GENERATIONS

UHMW polyethylene is a remarkable material that is extremely durable and has unusually low friction — that is, it is very slippery. Above are gears made from UHMW by Menasha's Poly Hi Solidur Group. At left is Poly Hi's fabrication facility in Oregon.

woodenware operations to concentrate exclusively on plastics, opening plants in Monticello and Manchester, Iowa, and later in Urbana, Ohio.

With Menasha's support, G. B. Lewis developed a custom molding business which later became the Menasha Molded Products Division. This unit made golf cart hoods, basketball backboards and other fiberglass-reinforced components for such companies as General Motors, John Deere, Harley-Davidson and Sears Roebuck.

In 1975, Menasha acquired full ownership of Lewis and renamed it LEWISystems. In a related initiative, two years before acquiring full ownership of Lewis, Menasha established a plastic pallet business through internal development under the trade name Convoy. This business, which serves a customer base similar to that of Lewis, has been extremely successful. Convoy pallets are widely used today in the food, beverage, grocery, automotive, textile and general material handling industries.

The Story of Poly Hi Solidur

Menasha's early investment in G. B. Lewis was a first step. Other plastics-related acquisitions and product development programs followed.

In 1971, Menasha acquired a 33 percent interest in a tiny, year-old company, Poly Hi Inc. of Fort Wayne, Indiana. Poly Hi processed and marketed an amazing new plastic, ultra-high molecular weight (UHMW) polyethylene, and was supplying UHMW sheet to Menasha's Appleton Manufacturing Division. Poly Hi's sales in 1971 were a mere $311,000. Over the next several years, as Poly Hi grew, Menasha kept increasing its investment, acquiring full ownership in 1977.

That acquisition was the genesis of today's Poly Hi Solidur Group, which has worldwide annual sales (including those of joint ventures) of more than $140 million.

UHMW, the main material processed and marketed by Poly Hi, is exceptionally durable and extremely "slippery." It is used, for instance, to line dump truck beds, to make artificial hip and knee joints and to manufacture critical components, such as gears and sprockets, in power transmission equipment. Another early application was Astro Ice, a plastic substitute for real ice in skating rinks. "Skating on Astro Ice is not as fast as regular ice," a Menasha Corporation publication noted, "but it can be used anywhere, anytime. Even traveling ice shows carry it with them for professional performances." Still another of UHMW's many uses is for the runners on dog sleds at the South Pole Station on Antarctica. If a more resistant material were used for the runners, the dogs would have to pull harder and would tire more quickly.

While Poly Hi was very good at finding uses for the new plastic and devising innovative production technologies, its production costs were high. Consequently, once an application was developed, lower-cost producers of UHMW often stepped in and captured the market because they were able to sell at cheaper prices.

Similar to the fellow who liked the electric razor so much he bought the company, Menasha eventually bought one of Poly Hi's largest and most efficient competitors, Scranton Plastics Laminating Corporation of Scranton, Pennsylvania, in this way obtaining the low-cost production capacity needed by Poly Hi to prosper in the UHMW business. Scranton Plastics was owned and managed by an entrepreneur named Del Kiesling. All five of his children worked in the business, and he was reluctant to sell. The delicate negotiations to acquire the company were handled by Tad Shepard, who spent many hours with the Kiesling family to convince them that Menasha would be a responsible owner. The acquisition, completed in 1981, was a major step in Poly Hi's growth. To this day, Poly Hi Solidur manufactures UHMW sheet at the plants acquired from Scranton Plastics.

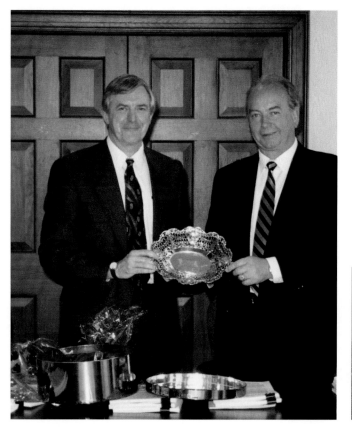

In 1997, when Menasha and Solidur Deutschland GmbH completed the merger of their UHMW polyethylene businesses, Solidur's Guenther Pennekamp, right, celebrated the occasion by presenting a tray to Bob Bero, Menasha's president and CEO. Below, Poly Hi Solidur is Menasha's most global business, as indicated by the flags outside the office of its South African subsidiary.

Continuing its growth, in 1985 Poly Hi expanded to Asia, forming a joint venture with Tsutsunaka Plastic Industry Co. Ltd. in Japan. Two years later, it expanded to Europe with the startup of a plant in Scunthorpe, England.

Meanwhile, a German company, Solidur Deutschland GmbH, had emerged as the leading European producer of UHMW sheet and, like Poly Hi, was entering global markets. Solidur Deutschland's owner was Guenther Pennekamp, an entrepreneur who had begun his career as an employee of the first company in the world to use UHMW resin in the manufacture of

AN ODYSSEY OF FIVE GENERATIONS

Originally a manufacturer of serving trays for the fast-food industry, the Traex Division has become an international supplier of more than 500 food service items.

products. Recognizing the enormous market potential for the material, he quit his job in 1965 and started Solidur Deutschland. In 1984, Bob Bero and Pennekamp discussed possible alliances between Menasha and Solidur, but nothing came of those talks. By 1992, however, Pennekamp realized he needed access to greater capital to keep expanding. Looking at the progress being made by Poly Hi and remembering his earlier conversations with Bero, he reopened the discussions.

Out of those negotiations came the transactions that propelled Poly Hi into the global leadership position it holds today. Those transactions were Menasha's 1993 purchase of the U.S. operations of Solidur Deutschland and its 1997 purchase of Solidur's operations outside the United States. Menasha owns 80 percent of the merged company, Poly Hi Solidur, while Pennekamp owns 20 percent and assists with high-level sales and strategic direction. "Guenther Pennekamp is a very good partner to work with," says Dale Mosier, group president, Poly Hi Solidur. "He is positive in his outlook and has been extremely helpful as we have learned to manage a global business."

Poly Hi Solidur is the premier supplier of UHMW both in the United States and internationally, with operations in nine countries. Its goals are to continue to expand globally and to invest in new products, new applications for existing products and the ongoing improvement of its manufacturing processes.

Targeting More Rapid Growth

In the 1970s and 1980s, as it grew in plastics, Menasha hired a number of experienced executives for this business and became more systematic in its approach to the market.

In 1977, Don Riviere was assigned to the Plastics Division to identify expansion opportunities. Two years later, Bero was recruited from Amsted Industries to head the division.

Division sales were $35 million when Bero came on board, with a goal of $65 million in three years. Later, the company established a goal of $100 million of sales from plastics by 1985. It achieved both goals and, since then, has continued to grow, reaching sales from plastics of more than $360 million in 1998.

In 1991, when the Plastics Division became a full-fledged group (subsequently it evolved into three groups), a company publication gave this description: "The Plastics Group is best characterized as a collection of highly profitable businesses which serve separate, individual market niches. Consolidation of the various plastics processes and products has never been a Group goal. By maintaining separate divisions, each business is allowed to focus intently on specialized customer needs."

Polymer Technologies Group

One way Menasha grew in plastics during the 1980s was by acquiring the companies that today form the Polymer Technologies Group. Having acquired these companies, Menasha invested in their continued development and in new products and new processes.

The first of these acquisitions, in 1984, was Traex Corporation of Dane, Wisconsin, now the Menasha Traex Division. Traex was founded in 1977 by a 29-year-old-entrepreneur, Greg Wenkman, a onetime corporate pilot who says he had "a deep desire to be in business for myself" and to ride in the back of the plane, not up front in the pilot's seat. With a $2,000 initial investment, he began operations in an old garage at the Madison, Wisconsin, airport, making plastic trays for fast-food companies. Traex grew rapidly and profitably based on low prices, consistent quality and rapid response to customer orders.

Skeptical that plastic trays were a growth business, but nonetheless looking for acquisition candidates in plastics, Tad Shepard and Don Riviere visited Traex in 1984. As Riviere recalled that meeting, "We talked with Greg and reviewed his company's financial statements. The growth rate and profitability were spectacular. We didn't say a word. But when Greg went to the men's room, Tad turned to me and joked, 'Did you bring the checkbook?' I said, 'No, did you bring it?' We were ready to buy the company right on the spot." Menasha ended up paying $3.5 million for Traex. Following the acquisition, Traex kept right on growing, with its sales doubling over the next three years.

Meanwhile, in 1987, Menasha purchased the Dripcut product line from another company to broaden the Traex product line. Dripcut's products included the familiar glass syrup pitcher, with a sliding valve as part of the handle, found in restaurants around the world. Dripcut also made plastic products, including salt and pepper shakers, sugar dispensers and cheese shakers.

By combining Traex and Dripcut, and continuing to invest in their growth, Menasha has built a premier business in the restaurant supply industry. Today, the Traex Division markets not only trays, but also plastic ketchup and mustard bottles, beverage pitchers and nearly 500 other items for food service companies in the United States and abroad. When you eat at Burger King, McDonald's, Hardee's or any other food chain, there's a good chance that part of the equipment needed to serve you — including trays, straw dispensers, French fry scoops and tumblers — was made by Menasha's Traex Division.

Also, like many other Menasha divisions, Traex is increasing its use of recycled materials to help protect the environment. Thermoplastics lend themselves to recycling. Across the nation, communities are collecting plastic products which are sorted by plastic type and delivered to reclaiming facilities for processing into pellets for sale to fabricators like Traex. For Traex, using recycled plastics is not only good for the environment, but is also good for business. Many of Traex's customers, themselves concerned with the environment, prefer trays and other products containing recycled resin. Nearly all Traex products now

PLASTICS

The Murfin Division makes an array of flexible screen printed products. It began by making the first plastic emblems for American Automobile Association members to attach to their cars and has continued with products such as these decals being made for a lawnmower manufacturer. In addition, it has expanded into high-tech areas such as control panel overlays and membrane switches.

contain at least some recycled material.

Two years after buying Traex, Menasha purchased Murfin, Inc., a Columbus, Ohio, printer of label and identity products. Murfin was founded in 1938 by Clif Murfin to make screen-printed plastic emblems for American Automobile Association members to attach to their cars. He launched his company in the belief that the heavy metal AAA emblems then in use were too expensive and could be produced at less cost with plastics. He was right, and the new AAA emblem won him a national magazine award for the first successful adaptation of injection-molded plastic for outdoor use.

From that start, Murfin, Inc., expanded into an array of printed plastic products. Today, as the Murfin Division of Menasha, it makes control panel overlays and membrane switches for microwave ovens, calculators and other products, and also manufactures nameplates, electroluminescence lighting and decorative trim. Its products include the battery tester on Duracell packages and circuits for blood glucose monitoring, as well as overlays and membrane switches for household appliances.

Seeking to broaden further its plastics business, Menasha spent four years trying to find a precision injection molding company. That search paid off in 1988 with the acquisition of the Thermotech Division of FL Industries. Founded in the late 1940s, Thermotech had been through a series of owners — ITT Corporation in 1970 and FL Industries in 1985 — before finding a home at Menasha. With operations in Minnesota, Florida and Texas, Thermotech is a leader in the precision injection molding of thermoplastic and thermoset engineered resins. It makes high-performance plastic components for companies in the automotive, medical, electronics and other businesses. Examples include door-lock panels for cars and gears for pumps.

Traex, Murfin and Thermotech — together with Montec Plastics, a Thermotech plant that became a separate operating unit in 1998 and that manufactures custom-molded products — are the four Polymer Technologies divisions. Jerry Roovers, group president, notes that each division is managed autonomously and has developed its own business strategy based on opportunities and customer needs in its particular market. Thermotech, for

AN ODYSSEY OF FIVE GENERATIONS

The Montec Plastics Division, originally part of the Thermotech Division, became a separate operating unit in 1998. At its plant in Monson, Massachusetts, above and left, the division makes a diverse line of custom-molded products including medical items, calculator cases and business machine components.

instance, is seeking to grow by offering customers greater design and assembly content, not just components. One of Murfin's strategies is to expand in the manufacture of polymer thick film (PTF) flexible circuits for use in medical products, such as skin-patch delivery devices.

Summarizing, Roovers says, "As a group, we are moving up the 'value chain' from a general plastics orientation to a focus on market segments where technical demands are greater and profit potential is higher."

Material Handling: The Birth of ORBIS

And what became of G. B. Lewis, Menasha's earliest plastics business? It has grown nicely and become part of the ORBIS Division, Menasha's main operating unit in the Material Handling Group. Orbis is the Latin word for loop or circle, which is appropriate for the division since it serves industry with returnable packaging that travels a continuous loop from supplier to factory and back for repeated use.

Many companies today insist on returnable, reusable packaging, such as for the shipment of parts from suppliers, for environmental reasons. They insist also that returnable packaging be cost-effective and save money.

ORBIS is Menasha Corporation's initiative to capitalize on these trends. It was created in 1996 by combining LEWISystems, the Convoy Plastic Pallets Division, and the DuraPAK and Donray businesses. Each of these units already made returnable packaging, such as heavy-duty plastic containers, pallets and protective interiors. ORBIS leverages their expertise by combining the units into a single entity focused on the returnable packaging market.

In addition, working with the Menasha Services Division, ORBIS provides complete returnable packaging solutions for companies that want to outsource their programs. This involves program design and management, including the inspection, sorting, repairing, storing and tracking of a company's containers and their return to suppliers for reuse.

Shortly before the formation of ORBIS was announced, Bob Bero said, "The desire to reduce packaging waste has created a huge opportunity for us. We are well positioned because we have the products, the plants and the distribution channels to supply durable plastic boxes. We are now building what we call a solutions business in returnable packaging. It not only involves being able to make the boxes and pallets, but also to manage a customer's inventory of boxes. Our thinking is shifting away from simply saying, 'Let's make good plastic boxes,' to, 'What does the market need?'"

The Tragic Deaths of Don Riviere and John Snyder

In January 1997, just six months after ORBIS was launched, tragedy struck when John Snyder, group president, Material Handling, and Don Riviere, Menasha's corporate vice president-strategic development,

AN ODYSSEY OF FIVE GENERATIONS

Many companies today insist on returnable, reusable packaging for the shipment of parts from suppliers, both for environmental reasons and to reduce packaging costs. Menasha's ORBIS Division serves this need with durable, reusable plastic containers and with complete logistical services to help companies manage their returnable packaging programs.

died in a private-plane accident.

The two Menasha executives were both licensed pilots and were flying together near Detroit when their small plane crashed. The deaths of these two well-liked men, both in the prime of their careers, was tragic for the men, their families, their many friends at Menasha and the corporation itself.

Riviere had joined Menasha in 1970 from DuPont and played a major role in establishing the strategic direction that guided Menasha to rapid growth in the 1980s and 1990s and continues to guide the company today. Snyder had joined the company in 1993 as vice president of the Material Handling Group, having previously headed his own company, Alexander Art Corporation in Salem, Oregon.

Bill Shepard, group president, Menasha Packaging, wrote, "These two men were simply outstanding. They gave us everything. As friends and associates, hopefully we gave back. We will miss them forever."

ORBIS Today

No company recovers easily from the deaths of two key executives. Menasha assigned Dave Rust, corporate vice president-human resources, to take temporary charge of the Material Handling Group and its ORBIS initiative until a new group president could be appointed. In late 1997, Mark Cane, president of Amtrak's Intercity Business Group and previously an executive at Burlington Northern Railroad, joined Menasha in that position.

Cane took over ORBIS at a time when it was still in its startup phase. Revenues were rising at a rate of about 20 percent a year, but earnings were lagging. "Merging all these businesses to form ORBIS has been challenging," Cane acknowledges. "Customers love us. We are making good progress in the marketplace and getting a lot of business. However, we must achieve satisfactory earnings to have wins for our shareholders and employees as well as for customers. I am convinced we are on the right track and this will happen in time." One of the questions is how to charge for container management and servicing in a way that makes sense for the customer and for Menasha, an issue that is being resolved — as in any new business — by testing various approaches.

Initial customers include vehicle manufacturers, lettuce growers and pharmaceuticals distributors. An example is the PACCAR Inc. truck plant in Denton, Texas, where ORBIS provides heavy-duty plastic containers that are used to transport parts from PACCAR's suppliers to the factory. ORBIS provides a total solution. Menasha employees are stationed on-site to manage PACCAR'S container inventory and handle all the logistics of inspecting, repairing, storing and tracking them and returning them to suppliers for reuse. PACCAR saves money by not throwing away packaging, and it reduces waste disposal in the process — and it does so without having to worry about getting containers to the right place at the right time or keeping its container inventory in good condition, since Menasha handles that part of the program.

ORBIS customers include many of the major industrial, wholesaling and retailing companies in North America. Menasha is not the only company in the returnable packaging industry. However, its competitive edge lies in its ability not only to supply the packaging, but also to provide a cost-effective total solution.

Listening to the Customer

Apart from being an important business initiative in its own right, ORBIS is indicative of Menasha's fundamental strategic direction. All seven of Menasha's business groups are becoming more market-driven: they are working in partnership with customers to understand and meet their needs and become preferred suppliers, and, where appropriate, they are providing total solutions that combine products and services.

As to Menasha's plastics businesses, which have grown so rapidly for the past two decades, Bero says, "We are moving away from a process-oriented corporation, what we would call a bricks-and-mortar mentality, toward a market-driven corporation. We characterize our businesses less today in terms of wood, paper, plastics and printing and more in terms of the markets they serve. The old plastics business isn't one business any more. It is really operating in many different directions, driven by end-use markets that are particularly attractive for products that happen to be made out of plastics."

AN ODYSSEY OF FIVE GENERATIONS

CHAPTER TEN

Completing the Package: Printing, Promotion and Corrugated Boxes

In 1995, continuing its evolution, Menasha reorganized into the seven business groups — and the 37 operating divisions within those groups that form the company today. The purpose of the reorganization was to bring similar operations together and provide a greater market focus for all of Menasha's businesses.

Moreover, the reorganization made clear just how diversified Menasha had become. As recently as the early 1970s, Menasha was in three main businesses: the manufacture of corrugated boxes, the production of corrugating medium at the Otsego mill in Michigan and the sale of timber from the Pacific Northwest. Today, Menasha continues in those businesses, but has added several more, including its three plastics groups discussed in the prior chapter.

Menasha's unusually high degree of diversification, including the continued

Above is a package designer at the Menasha Art Center in Menomonee Falls, Wisconsin. Although each box plant has its own design staff, clients' more complex design needs are handled by the Art Center. At left is a sampling of printed materials from the Neenah Printing Division.

rapid growth of the Packaging Group (corrugated containers) and the creation of still another group, Promotional & Information Graphics, is purposeful. Diversification not only provides a balanced investment for Menasha shareholders, but also helps offset the tremendous cyclicality of Menasha's timber and corrugating medium businesses. This cyclicality has, if anything, increased in recent years.

The timber and corrugating medium businesses can be very profitable when market prices for the products are high, yet can suffer badly when prices falter. Dramatic shifts in profits often occur without warning. For instance, after earning a record $35 million in 1995, the Otsego mill plunged to an operating loss two years later as the market price of corrugating medium dropped 50 percent in just 12 months.

Bob Bero describes timber and the Otsego mill as "900-pound gorillas — they

Menasha entered the printing business in the 1980s as a diversification move. The Fox River Valley of Wisconsin is a traditional center of high-quality commercial printing.

Commercial Printing

One portion of Promotional & Information Graphics is commercial printing. Menasha is one of several high-quality printing firms in the Fox River Valley, a center of commercial printing in the United States.

Menasha entered this business in 1986 when it purchased Neenah Printing and its subsidiary, Oshkosh Printers, Inc. Menasha did not have a planned strategy to go into printing, although of course the business does tie closely to other parts of the company, including corrugated boxes which typically are printed before being shipped to the customer.

"Printing was another chance to diversify," Tad Shepard explains. Neenah Printing is a highly respected local company founded in 1900. Shepard knew the owners, David Thomsen and Norman Brown, and Menasha bought the business when it became available at a favorable price. In that sense, the acquisition was opportunistic as much as strategic.

Neenah Printing, which provides a full range of services in commercial, business forms and packaging applications, has continued to expand under Menasha's ownership. In 1988, for instance, it completed a $2 million facility for the manufacture of business forms. With the equipment and technology investments made in recent years, Neenah Printing is today among the most advanced full-service printers in North America.

In 1993, Menasha acquired another

go where they want to go," meaning that Menasha has little or no control over the selling prices for their products. He asserts, "We need to offset the feast or famine of our timber and paperboard businesses."

The Promotional & Information Graphics and Packaging groups help Menasha do just that, while providing growth opportunities in their own right.

COMPLETING THE PACKAGE: PRINTING, PROMOTION AND CORRUGATED BOXES

Above is a sampling of printed products. At right is a facility in Promo Edge's fulfillment business — involving the printing and warehousing of promotional items and other products for customers and filling orders for those products.

printer, New Jersey Packaging of Fairfield, New Jersey, which specializes in pressure-sensitive and heat-seal labels for the pharmaceutical and health care industries.

Printing is a highly competitive industry, yet one in which Menasha continues to perform well. "We can be effective operators in commercial printing and make attractive returns," Bero says. "On top of that, if we can grow the business at a rate of 10 percent a year, we will be very happy."

Menasha's Strong Position in Promotional Graphics

The other segment of the Promotional & Information Graphics Group is Promo Edge, a national leader in the design and manufacture of in-store promotional materials for consumer product companies. This segment links closely with commercial printing in that it involves the printing of specialized materials such as temporary in-store displays, product labels, on-package coupons, promotional games, "shelf talkers" and signage. These materials are a fundamental tool used by tens of thousands of companies to market their products, make them stand out from competing products and stimulate sales.

Promo Edge serves a national roster of blue-chip customers, including General Mills, Kraft, Pillsbury, S. C. Johnson Wax and Gillette, among many others.

Development of the Color Division

Although Promo Edge is a relatively new division formed in 1996, Menasha Corporation's roots in promotional graphics date back to 1972. The resourcefulness of Menasha people in developing the promotional graphics business and keeping pace with evolving customer needs, culminating in the formation of Promo Edge, is an example of American enterprise at its best.

Menasha got into the business as an offshoot of its corrugated box operations. In the early 1970s, facing intense competition in corrugated boxes from large national competitors such as International Paper and Container Corporation of America, Menasha began to look for ways to differentiate its box business through service and product specialization. In its search for profitable niches, Menasha developed the capability to print high-quality color on corrugated boxes.

At that time, most other box companies treated high-quality color printing as if it were not worth the bother. However, packaging was changing, and Menasha recognized this fact. The emergence of discount retail chains and self-service stores, such as warehouse clubs, meant that many products were now being displayed on store shelves in their original corrugated shipping containers. The traditional brown corrugated box with black printing was suddenly being called on to display and sell products.

In 1972, Menasha purchased a corrugated box plant in Hartford, Wisconsin, and

In-store promotional graphics (the temporary product displays found in supermarkets, pharmacies and other retail stores) are a large and growing business.

Menasha's Promo Edge Division, which produced the displays shown here, is a single-source supplier of promotional graphics and related products and services.

formed a Color Division. "The Hartford plant, when we acquired it, was not much different than any other box plant," Tad Shepard recalls. "However, we bought specialized printing and die-cutting equipment and put it into Hartford, developing the plant as our first color location." This equipment allowed Menasha to print letters, designs, drawings and/or photographs — whatever the customer wanted — on boxes in color.

In announcing the new venture, Menasha's employee publication, the *Log*, proclaimed that "full service is the corporation's intent and promise." Explaining what full service meant, the publication added, "Yes, we structure the box: design its intricate die cuts, slots, folds and closures and produce them. We print, we laminate, we label, we coat to prevent abrasion. And all in all, we make a beautiful box which protects and sells." Besides producing the box, Menasha developed a comprehensive graphic design capability to help customers create "rich and dazzling" boxes to stand out from other boxes on the store shelf and shout "buy me" to the retail customer, in the words of the article.

The new division was successful from the start. Menasha expanded by opening color-printed box plants in Olive Branch, Mississippi, and South Brunswick, New Jersey, in addition to the plant in Hartford, Wisconsin. By the late 1980s, however, many other box companies had entered the color-printed business and profit margins had narrowed. Seeking a new niche where it could apply its skills in making complex color-printed products from corrugated sheet, the Color Division began designing and manufacturing point-of-purchase displays. It was this transition that brought Menasha into the heart of the promotional graphics business. For Anheuser-Busch, for instance, Menasha produced an in-store display in the form of a sleek race car — made from color-printed corrugated and filled with cases of Bud Light beer for sale. For Minnesota Mining & Manufacturing Company, Menasha produced a 3M Active Strip display that won a coveted gold award from the Point of Purchase Advertising Institute.

With its focus on creativity, product quality and customer service, the Color Division increased its point-of-purchase display volume by 45 percent annually in the three years from 1991 through 1993. By early 1994, point-of-purchase displays constituted about half the division's total sales volume, with prospects for continued growth. As a result, in February of that year the Color Division changed its name to DisplayOne.

Menasha originally entered the color-printing-on-corrugated-board business with the formation of its Color Division, a predecessor of Promo Edge. The Color Division produced innovative packaging and display designs, such as this in-store display for Anheuser-Busch.

The Acquisition of Mid America

Meanwhile, in 1985, Menasha bought one of the nation's leading makers of merchandising tags and labels, Mid America Tag & Label Co. Mid America's founders were Bob Dunsirn, who ran the manufacturing and engineering side of the business, and Don Buchta, who was in charge of sales and marketing. They began operations in 1966 in a small building in downtown Neenah. Their idea was to provide consumer and industrial product companies with total service, from concept to finished product, in tags and labels, such as the labels on pickle jars or the decals sometimes attached to soda bottles.

Through persistence, intensive customer service and investments in technology, Mid America steadily captured market share. Within two decades, it was supplying tags and labels to such companies as Coca-Cola, Pepsi, Briggs & Stratton, Oscar Mayer and Sheaffer (pens).

The company's engineers pioneered a special 14-color flexographic press to produce high-resolution labels in rolls for automatic application. Mid America also screen-printed decals, product name plates and decorative trims on a variety of papers, plastics, films, foils and adhesives. When acquired by Menasha in 1985, Mid America was in the process of starting up a second 14-color press. The new unit was introduced with great fanfare. A Menasha Corporation publicity brochure proclaimed that the press produced "glowing labels, brilliant decals, vivid multi-layer coupons." Promotional games printed on the press, the brochure declared, "spring into life." Mid America even designed a special logo, with the words "Fourteen Colors," to celebrate the occasion. Conveniently, the words "fourteen colors" have 14 letters, so that each letter in the logo could be printed in a different color, driving home the message of the new equipment's capabilities.

The acquisition of Mid America was a big step for Menasha into the promotional graphics business, providing a vehicle for growth. Since completing the acquisition, Menasha has continued to expand the Mid America Division's operations and sales by investing in people, equipment and customer service programs.

In addition, Mid America's tag and label business serving industrial customers was split off from the consumer products segment of the business and became the Printed Systems Division of Menasha. This division makes bar-coded inventory-control tags and other products for use in manufacturing, warehousing, distribution, transportation, and repair and rental facilities. An example is the familiar coded tags, showing flight number and airport of destination, that airlines attach to the handles of checked luggage.

In 1989, Menasha acquired Labelcraft Corporation, which became the Farmingdale, New Jersey, plant of the Promotional & Information Graphics Group.

A Single-Source Supplier

As the promotional graphics business continued to evolve, in 1996 Menasha merged DisplayOne, Mid America and its point-of-sale business to create a new division, Promo Edge, which has revenues of approximately $60 million a year. As is the case with ORBIS in the Material Handling Group, Promo Edge is an example of Menasha's determination to achieve greater customer focus and deliver total solutions that combine products and related services.

Bob Bero notes that the design and manufacture of in-store promotional graphics is a multi-billion-dollar-a-year industry that is very fragmented with no dominant market leader. The business is essentially a printing industry that is highly sensitive to price, quality and customer service.

Through Promo Edge, Menasha wants to become the undisputed market leader by being the first company to provide not only one-stop shopping for a full line of products and an exceptionally high level of service, but also by offering related research — in other words, by providing the whole package of products and services needed by a consumer products company.

"The basic objective," Bero explains, "is to take the variety of printing capabilities we have in point-of-purchase displays, coupons, labels and collateral materials, including materials from our commercial printing plants, and combine them with products from other printing companies — things we don't make — to deliver a comprehensive array of products and services."

COMPLETING THE PACKAGE: PRINTING, PROMOTION AND CORRUGATED BOXES

Printed Systems makes bar-coded inventory-control tags and other products for use in manufacturing, warehousing, distribution, transportation, and repair and rental facilities.

Menasha's Solid Fibre Division makes an unusual product — solid fiber board that is durable, economical, and moisture- and puncture-resistant. These rugged UPS tote boxes are a typical application.

Services include product design and the management of promotions from start to finish. Promo Edge also offers market research to provide customers with knowledge of retailers' promotion requirements and information on consumer motivation. "We can identify special needs Wal-Mart might have, for instance, in running promotions," Bero states. "We can then work with our customers — the companies supplying products to Wal-Mart — to meet those needs."

Marketing and promotions executives in corporations across the U.S. are working harder than ever, and their time is stretched thin. Consequently, many consumer products companies are outsourcing some logistical and management aspects of their in-store promotional activities and are working with fewer core suppliers. With its Promo Edge program, Menasha seeks to capitalize on these trends. In the case of Kraft, for instance, Menasha is now the exclusive supplier of certain materials, and two Menasha employees have offices on-site at Kraft to ensure close coordination with that company's marketing and sales personnel.

Bero is optimistic that the Promo Edge Division will become a much bigger business that is recognized as the premier supplier of in-store promotional graphics to consumer products companies nationwide.

Menasha Packaging

Newer businesses like Promo Edge and ORBIS are important to Menasha's future. One of Menasha's most aggressive growth

The labels in the upper photograph were made by the New Jersey Packaging Division, which serves the pharmaceutical industry. The lower photograph features labels made by Promo Edge.

programs, however, involves its largest business — corrugated boxes, a "mature" business Menasha has been in since 1927.

Since the 1970s, Menasha has carved out a strategy of competing by providing a high level of service and filling special orders, including short-run orders. "We continue to make good money in corrugated packaging," Bill Shepard, group president, Packaging, says. "Why are we doing so well in a low-return industry? The reason is that we have a template from which we will not deviate: we sell high-quality products and provide creative design, outstanding graphics and incredible service."

"Our attitude," Shepard observes, "is that our template works, so we ought to apply it in more markets." Consequently, Menasha is now applying its success formula to expand to additional markets. As of 1996, when the growth strategy was in its early stages, Menasha was producing and supplying corrugated boxes in five regional markets across the United States. Bill Shepard says that by building and acquiring additional corrugating and converting (box) plants, the company intends to be in 11 regional markets early in the new century.

From mid-1995 through early 1998 alone, in implementing the strategy, Menasha built corrugating or converting plants in Erie, Pennsylvania, and St. Cloud, Minnesota, and acquired plants in Cullman, Alabama; Phoenix, Arizona; and Middlefield, Ohio.

If all goes according to plan, and so far the strategy is proceeding smoothly, the

AN ODYSSEY OF FIVE GENERATIONS

COMPLETING THE PACKAGE: PRINTING, PROMOTION AND CORRUGATED BOXES

These photos, left and above, feature a sampling of corrugated containers made at Menasha's 13 box plants across the United States. Although corrugated boxes are considered a mature business with a high level of competition, Menasha's Packaging Group has succeeded and grown faster than the industry by emphasizing quality products, creative design and a high level of customer service.

Packaging Group's sales will increase eight-fold by the year 2010. It would doubtless stun and thrill Mowry Smith, Sr., to know that Menasha's first box plant not only saved the company from bankruptcy, but on top of that has evolved into one of the company's most powerful engines for growth in the twenty-first century.

Solid Fibre and Sus-Rap

Two other divisions in the Packaging Group, Solid Fibre and Sus-Rap, have their own unique and interesting stories.

The Solid Fibre Division grew out of Wisconsin Container Corporation, founded in the city of Menasha in 1928 by John Strange Paper Company, Menasha Corporation and several other local box manufacturers. A relatively small operation with a narrow product focus, Wisconsin Container was created to make solid fiber laminated paperboard, which competed with corrugated in the 1920s and 1930s in the manufacture of shipping boxes. Laminated paperboard contains anywhere from two to five plies of board, depending on the thickness and strength desired, bonded into a single sheet. Unlike corrugated, it does not have a squiggly center.

After World War II, the use of laminated paperboard shipping boxes faded because of corrugated's lower cost and lighter weight. Nonetheless, Wisconsin Container continued to generate strong profits by developing specialized applications that capitalized on the board's durability as well as its moisture and puncture resistance. These applications included bottoms for fiber drums and protective packaging for steel coils. "The interesting part of Wisconsin Container is that money can be made even in a declining business by being a low-cost producer and being innovative," Tad Shepard relates.

Menasha purchased John Strange and its ownership interest in Wisconsin Container in 1969, acquiring full ownership of Wisconsin Container one year later. The unit became Menasha Corporation's Fibre Products Group, later renamed the Solid Fibre Division. Under Menasha's ownership, the division continued to prosper by devel-

William Shepard, a fifth-generation Smith, is president of the Packaging Group, Menasha Corporation's largest business. He has been a Menasha Corporation employee since 1983 and was elected to the company's board of directors in 1998. Prior to joining Menasha, he was a project engineer at International Paper.

oping new applications for laminated paperboard, including containers for hot-melt adhesives. Not only are these containers leak-proof, but their interiors are coated with silicone so the hot-melt adhesive doesn't stick to the packaging. Walter Sellnow, Wisconsin Container's president who joined Menasha following the 1970 acquisition, recalls, "Although we weren't making a glamorous product, Menasha gave us room to innovate and grow. At one point, Tad said, 'I don't know what you're doing. But you're making money. Keep it up.'"

Laminated paperboard is strong and durable, is less expensive than plastic and lasts longer than corrugated. Its major use today is for material handling containers and protective packaging. The Solid Fibre Division continues to operate at the original Wisconsin Container Corporation plant, modernized in recent years. The division's products include its patented containers for hot-melt adhesives, tote boxes (such as those used by United Parcel Service to move shipments within its terminals) and pallets.

Like Solid Fibre, the Sus-Rap Division is a specialty packaging business that generates good profits and cash flow. The division's origins date back to the early 1950s when a

COMPLETING THE PACKAGE: PRINTING, PROMOTION AND CORRUGATED BOXES

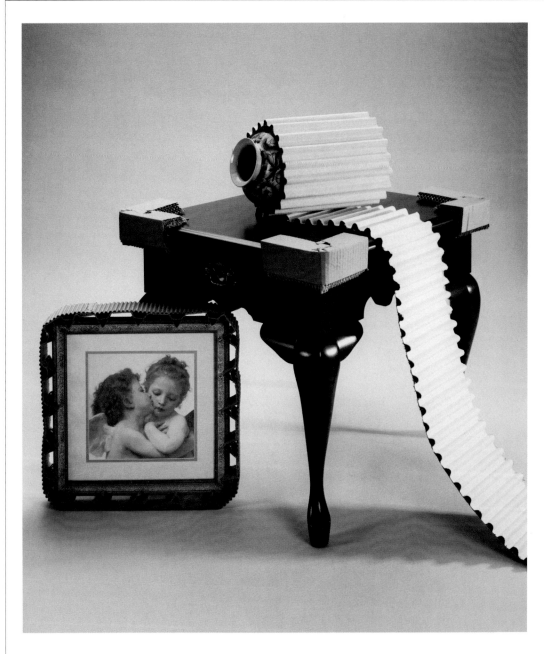

Menasha's Sus-Rap packaging material is named for its ability to "suspend" a product inside a shipping container. When a product, such as a mirror or electronic gear, is encased with Sus-Rap, it is supported and cushioned by the projections and grooves.

Milwaukee inventor, Martin Van Antwerpen, designed a machine to manufacture a unique "suspension wrap" (hence the name Sus-Rap) packaging for replacement windshields. This packaging was wrapped around the outer edge of the windshield to suspend it inside a corrugated container, protecting against breakage during shipment. The packaging, made from composition board, was produced in a continuous strip several inches wide. Grooved "fingers" protruding from the strip held the windshield firmly in place inside the container.

Menasha acquired Van Antwerpen's company, Vanant Packaging Corporation, in 1968 and continued to develop the product, adding new variations and new materials, including "poly Sus-Rap" with a satin-smooth laminated finish to protect surfaces prone to abrasion. Today, the Sus-Rap product is widely used to hold and protect a variety of fragile products inside boxes, including electronic equipment, mirrors, framed pictures and clocks. In addition, the division makes corrugated Menasha Pads used by the furniture industry to cushion and protect the corners and sides of furniture in transit.

Packaging is a huge business with constant change and innovation, and Solid Fibre and Sus-Rap are examples. Neither is especially large. However, they are very good businesses that highlight one of Menasha Corporation's strategies: find niche markets where the company, with its strong customer focus and ability to innovate, can grow profitably.

CHAPTER ELEVEN

Menasha Corporation Today: Something to Be Proud Of

Change, a constant in business, is always creating challenges. New products come. Old products go. Customer needs evolve. Technologies make leapfrog advances. Cost pressures intensify.

Few companies can survive this torrent of change for as long as a century. Even the largest and most prosperous corporations often succumb to change and go out of business or are acquired. Who knows today that American Cotton Oil Corporation, Distilling & Cattle Feeding Company and United States Leather Company, mere footnotes to history, were three of the 12 original stocks in the Dow Jones Industrial Average in 1896? In fact, of the 12 large blue-chip companies that originally made up the average — many in businesses that, like woodenware, went the way of the buggy whip — only General Electric Company remains in the index today.

Menasha Corporation has survived and grown not just for a century, but for a

Robert Bero, above, was named Menasha Corporation's president and chief executive officer in 1993. Left, fifth- and sixth-generation Smiths gather at the annual family picnic in Smith Park.

century and a half. It has done so because of its ability to change at critical junctures and because of the unflagging support of its Smith family owners. Bob Conrad, a longtime Menasha Corporation salesman and sales manager who retired in 1992, remembers phoning Tad Shepard in the 1980s when he heard a rumor that Menasha was about to be sold. "Tad's answer was a classic," Conrad recalls. "He told me, 'Before Menasha is sold, I will buy the interested party.'"

Menasha's ability to keep pace with the times and renew itself at key turning points is made clear by its roster of products. Not one of the products made by the company in 1849 is still made by the company today. Correspondingly, not one of the products it makes today was made by the company in 1849.

Yet, in other ways, Menasha has remained amazingly the same. The values established by founder Elisha D. Smith, including integrity and concern for people, continue to guide the company as they have for 150 years. CEO Bob Bero says, "There are two things I hear from people on a regular basis. One is, 'We have never seen a private company that is 150 years old where the shareholders are still so supportive.' The other is, 'Are you guys for real? Are you really as honest and truthful as people say?'"

Bero points out that integrity and concern for people, besides being the right things to do, are effective business strategies because they give Menasha staying power. Employees like to work for a company that cares about people and customers like to do business with a company that is honest and fair in its practices. From the day Elisha Smith bought the Pail Factory in 1852, Menasha has been managed with a goal of creating long-term value for customers, employees and shareholders, not with a purpose of making a quick buck.

Menasha director Richard Clarke asserts, "Menasha represents what a lot of people think America represents. It embodies the values that have made our country great."

Expanding in a Competitive Economy

Throughout its history, the Smith family shareholders have been loyal and supportive, sticking with their company even when it provided them with virtually no cash dividends.

This loyalty has been well rewarded since the early 1980s, when Menasha Corporation's management and board of directors committed themselves, through the adoption of a Mission Statement, to "provide steady growth, reasonable and dependable dividends and a return on equity better than average for all U.S. companies."

The adoption of the Mission Statement was followed by a burst of profitable expansion. From 1982 to 1995, Menasha's revenues advanced from $195 million to $915 million; its earnings from $7 million to $62 million; and its return on equity from seven percent to 19 percent. As earnings rose, dividend payments increased from $20 per share in 1982 to $127 per share in 1995. The latter, which included $49 per share of special and liquidating dividends, represented a total distribution to shareholders of $21 million in 1995.

In leading the company to a dramatically higher level of financial performance, successive CEOs recognized it was no longer possible to generate low returns and survive as a company. The global business environment had become far too competitive for any company to underperform and endure.

Menasha has taken many risks and undergone many transformations in its 150-year history, including its entrance into the corrugated box business in 1927 and the building of the huge Anaheim box plant in 1955. This latest transition to a company focused more intently on its financial performance is another milestone: it is as vital as any change the company has ever undertaken.

But that transition was only the beginning.

The reality today is that competitive pressures keep getting more intense. The challenge faced by all companies is to establish and maintain leadership positions in core markets and leverage those positions to grow profitably. Family-owned businesses such as Menasha, while they have the distinct advantage of being able to think and act long-term, are not immune to these forces. Building leadership positions in core markets, whether in-store promotional graphics, color-printed corrugated boxes or returnable packaging, offers a way to grow. An inability to establish leadership positions is an almost certain route to failure.

In 1995, recognizing these various factors, CEO Bob Bero hired a prominent consulting firm, Strategic Decisions Group of Palo Alto, California, to help Menasha develop a more aggressive strategic plan and raise its financial sights another notch or two. Again, Menasha was ready to take on risk and undergo change. In announcing this planning initiative, Bero told employees, "Menasha Corporation recently embarked on a very important project that will impact every one of us: the development of more robust, value-enhancing strategies for many of our businesses." He also said, "Our mission is to construct a

Employee teamwork is a vital element of Menasha's operating philosophy. Pictured is a project review meeting at the Menasha Art Center.

AN ODYSSEY OF FIVE GENERATIONS

In 1996, CEO Bob Bero visited customers in Japan to emphasize Menasha's customer focus and learn more about their needs. Here he chats with officials of Tenryu Lumber Ltd., Nichimen Corporation and Pre-cut Sawmill, Tokyo. Menasha's growing emphasis on customer-focused services in both the U.S. and internationally is illustrated by the ORBIS Division, below left, and the Menasha Services Division employee repairing a customer's returnable containers, right.

group of businesses that in total perform better than the average performance of American businesses. To accomplish this, we must direct the human and investment capital of the company toward business opportunities that have better-than-average growth and return potential."

Out of this intensive, yearlong planning effort came a series of tightly defined business strategies and new ventures. A common theme across all these strategies and ventures was, and is, a greater focus on customers. Bero's vision is that Menasha will deliver such a consistently high level of value and service that it will become a preferred supplier to all its customers.

Two examples are the ORBIS reusable packaging business in the Material

MENASHA CORPORATION TODAY: SOMETHING TO BE PROUD OF

Bob Bero addresses attendees at Poly Hi Solidur's global conference. Although international markets currently account for only about 10 percent of the company's revenues, one of Menasha's objectives is to grow rapidly overseas.

Handling Group and the Promo Edge Division in the Promotional & Information Graphics Group, both of which grew out of the 1995 planning program. "ORBIS and Promo Edge are in the forefront of changes taking place in their markets," Bero asserts. "If they live up to our expectations, they will generate large payoffs for the company."

Another strategy developed in 1995 is the current program to grow rapidly in corrugated boxes by entering additional regional markets.

Still another key opportunity for Menasha Corporation is global expansion. International markets currently account for about 10 percent of the company's revenues. At present, the Poly Hi Solidur and Forest Products groups both have a significant market presence outside North America. In addition, since the early 1990s, the Traex Division has expanded its sales of plastic trays and other products to the fast-food industry in Europe and Asia. Bero vows, "We'll be more international in the future. We can't be the size company we're trying to be without international participation."

Ultimately, ORBIS, Promo Edge, growth in corrugated and greater penetration of global markets are just a few of the ways in which Menasha intends to expand. Bero seeks to increase the company's revenues at an annual rate of at least 10 percent and generate a long-term average return on equity of more than 15 percent, ambitious targets for any company. The most critical element in achieving these goals will be Menasha's ability to continuously reexamine and improve every one of its businesses. Indeed, each of the company's 37 divisions has developed a strategic plan to capitalize on opportunities in its markets and maximize its long-term financial performance.

Almost by definition, implementing a strategic program of this magnitude can cause temporary disruptions and penalize short-term earnings. To make matters more complicated for Menasha, in 1996 and 1997 — just as it was putting its new strategies in place — the company's two large cyclical businesses, paperboard and timber, went into earnings slumps when market prices for their products fell sharply. Moreover, Menasha began installing new SAP operating software at many of its divisions, and this required a period of learning and adjustment by employees.

Taking on all this change was expensive in the short run. Following 13 consecutive years of strong revenue growth, net sales were flat in 1996 and 1997, and earnings declined precipitously from their 1995 high.

But those problems now seem to be fading. As of mid-1998, Menasha's strategic plan was taking hold, and earnings were recovering. Bero remains confident that Menasha is resuming its pattern of revenue and earnings growth. He recently told employees, "Change is necessary in today's business world to maintain the strength of the company, but it can challenge the goal of constant earnings improvement. We are now entering a phase where we can put some of that change to work for us."

Of course, Menasha has been through change before. The current period of adjustment pales in comparison to the wrenching change encountered in the late 1920s when Menasha moved out of woodenware into corrugated. Menasha's history stands as testament to the inevitability of change and the fact that no company can stand still and survive.

Summing up, Bero says, "Here is a company that clearly is moving from a narrower perspective to a broader perspective, not only in terms of the diversity of its products and the professionalization of management, but also in terms of the sophistication of its business techniques. This is all part of a continuum, rooted in this absolutely incredible sense of heritage and the love of the owners for this company."

A Special Bond Between Family and Company

Menasha Corporation is surely one of the most unusual companies in America. It has evolved and thrived during a century-and-a-half of ownership by the same family. Reflecting the family's values, the company cares about increasing its earnings, but not at the expense of people. Elisha Smith not only built a successful business, but he treated people with respect, conducted his business and his personal life with integrity and went to exceptional lengths to make the world a better place. As it looks to the future, Menasha seeks to grow and deliver excellent returns to its shareholders while embracing and broadening its wonderful heritage.

The fourth- and fifth-generation Smiths who currently own Menasha Corporation hope that future generations will understand their company's rich history and will recognize the special bond that exists between family and company. They hope future generations will want to keep Menasha Corporation in family hands, maintaining the private ownership structure that each of the first five generations has worked so hard to preserve and strengthen.

As Mowry Smith, Jr., puts it, "Menasha Corporation is more than a business. It is an extension of the family. It is something of which we all are proud."

Above, Mowry Smith, Jr., right, chats with Walter Stommel, who joined Menasha in 1928 as an office boy, retiring 46 years later in 1974 as personnel manager of the Neenah container plant. Top, left to right, John Sensenbrenner, Jr., a fourth-generation Smith, and his wife, Mary, join Anne Des Marais, a fifth-generation Smith, and her daughter, Emily, at a ceremony honoring the site of the original Elisha D. Smith Library. The site is a stop on a walking tour of historic Menasha.

Sustaining Elisha's Generosity

Beginning with the extraordinary generosity of Elisha D. Smith in the nineteenth century, involvement in the community and financial support of worthy causes have become a tradition of the Smith family, Menasha Corporation and the company's employees.

Charles R. Smith carried forward many of his father's charitable programs, including his support of religious and educational groups. Charles's son, Mowry Smith, Sr., was, in turn, deeply involved throughout his life with the Boy Scouts and a host of other nonprofit organizations.

In 1953, when Mowry, Sr., was company president, the family decided to formalize its contributions program by creating the Charles R. Smith Foundation. After Menasha Corporation acquired John Strange Paper Company in 1969, it merged that company's foundation — the Hugh Strange Charitable Foundation — with the Charles R. Smith Foundation to form the Menasha Corporation Foundation, today the company's principal instrument of philanthropy.

One objective in channeling support through the foundation is to provide a consistent level of contributions. Until 1995, the company gave one percent of its pre-

Lydia Shepard, a sixth-generation Smith, visits the Fox Cities Children's Museum in Appleton, Wisconsin. The handprints on the wall — including Lydia's, on behalf of the Menasha Corporation Foundation — recognize the museum's many donors. The foundation supports the museum as part of its ongoing commitment to help enhance the quality of life in communities where Menasha Corporation has operations.

tax earnings to the foundation. That amount was increased at the urging of several family members. As a result, the company now gives the foundation an amount equal to 1.5 percent of its average pretax earnings for the prior three years. This averaging assures consistent funding even in years when Menasha Corporation's earnings are down. In 1997, the company gave the foundation a record $1.2 million, more than double the amount three years earlier.

One of the basic tenets of the family, the company and the foundation is that giving of oneself is as important as giving dollars. The foundation's 1997 donation to Habitat for Humanity is an example. Habitat for Humanity is a nonprofit organization that organizes donations and volunteers to build homes for families in need. The foundation became the main sponsor of the group's Greater Fox Cities project, donating $20,000. In conjunction with that gift, more than 100 Menasha Corporation employees, along with family and friends from the Neenah area, volunteered their time and expertise to build a home for a family.

Another goal is to identify pressing human needs and target donations for maximum impact. For instance, the foundation recently increased its support of organizations such as CAP Services that are helping Americans on welfare get into the workforce. "Many of these people will need training, child care and transportation assistance for a short period of time, and the government cannot and should not provide it all," says Steven S. Kromholz, the foundation's president.

The foundation seeks, in particular, to make a difference in communities where Menasha Corporation has operations. "We recognize a special obligation to these communities because we are part of them," Kromholz says. He notes that most Menasha plants are located in smaller cities and towns. "For instance," he says, "we have a plant in Coloma, Michigan, where we employ about 150 people in a community of 1,800. By providing jobs and donating money to local nonprofit groups recommended by our employees, we can have a positive impact. I think the fact that Menasha is located primarily in smaller communities, together with the way the foundation operates, has done a lot to create the tremendous employee loyalty and esprit de corps in this company."

The foundation occasionally reaches outside its plant communities to help nonprofit groups that have a critical financial need not being met by other donors. This includes causes of personal interest to members of the Smith family. A recent case involved a matching gift to New Dramatists, a New York nonprofit organization. New Dramatists is a national cultural resource, training young playwrights, actors and actresses. Faced with severe financial problems, it was able to recover and broaden its base of donors because of the matching gift provided by the foundation.

The board of directors of the Foundation are, sitting left to right: Donald C. (Tad) Shepard, John Sensenbrenner, Jr., Clark Smith and Thomas Prosser; standing, left to right: Peter Radford, Susan Gosin, Onnie Smith, James Sarosiek, Oliver Smith, S. Sylvia Shepard, Steven S. Kromholz, Karen Helgerson and Robert Bero.

Smith Family Tree

Generation... 1st	2nd	3rd	4th	5th	6th
(1827-1899) **Elisha Dickinson Smith*** **Julia Ann Mowry**	(1855-1916) Charles R. Smith* Jennie Mathewson* (Isabel Rogers* 2nd wife)	(1891-1964) Mowry Smith, Sr.* Katharine Ives*	Mowry Smith, Jr.* Mary Leach*	Mowry Smith III Curtis Smith Tam Smith Onnie Smith	Courtney Smith Carleton Mary Smith Molly Smith Oakley Smith Jesse Smith
			Carleton L. Smith* Josephine Kimberly*	Carleton K. Smith	Nick Gansner Andy Gansner Elliott Gansner
			Katharine L. Smith* Don Gosin*	Katharine Gosin Gansner William Gansner	William Barrett Jack Barrett Henry Barrett
				Susan Gosin Richard Barrett	Katie Goelz Linna Goelz Peter Goelz
				Jennie Gosin Goelz Richard Goelz	
		(1892-1973) Carlton R. Smith* Theda Peters*	Lawton Smith* Ann Ellis	Charles A. Smith	
			Carla Smith* William D. Radford	Paula E. Riggi Kavon Riggi	Kaylee Riggi Mariah Riggi
				Theda Jessen Huascar Jessen	Olivia Roup
				Peter Radford Teresa Radford	Carrie Radford David Radford
				Carlton Scott Radford	
				Curt Radford Kathy Radford	Carla Radford Weston Radford Quintin Radford Samuel Radford Carlton Radford
			Tamblin Smith		
			Oliver C. Smith Pat Pierce	India R. Clarke John Clarke	J. Taylor Clarke Chelsea Clarke Caroline Clarke
			Sylvia Smith James A. Vaccaro	Pierce Smith	
				Marc B. Vaccaro Kate Vaccaro	Kristina Vaccaro Alison Vaccaro Holly Katheryn Vaccaro
			Clark R. Smith Trina Hendershot	Todd Vaccaro	
	(1853-1854) Mary Smith*	(1895-1975) Sylvia Smith* Donald Carrington Shepard, Sr.*		Clark Carlton Smith Luke Owen Smith	
	(1872-1941) Jane Smith* S. Elmer Smith* (E.E. Haskin* 2nd husband)		D.C. Shepard, Jr. Jane Steinborg	S. Sylvia Shepard Will Holtzman	Katie Holtzman Josie Holtzman
				Donald C. Shepard III Jane Shepard	Christopher Shepard Lisa Shepard Peter Shepard
				William Shepard Cynthia Shepard	Lydia Shepard
				Julia Shepard Waite Michael Waite	Steven Waite

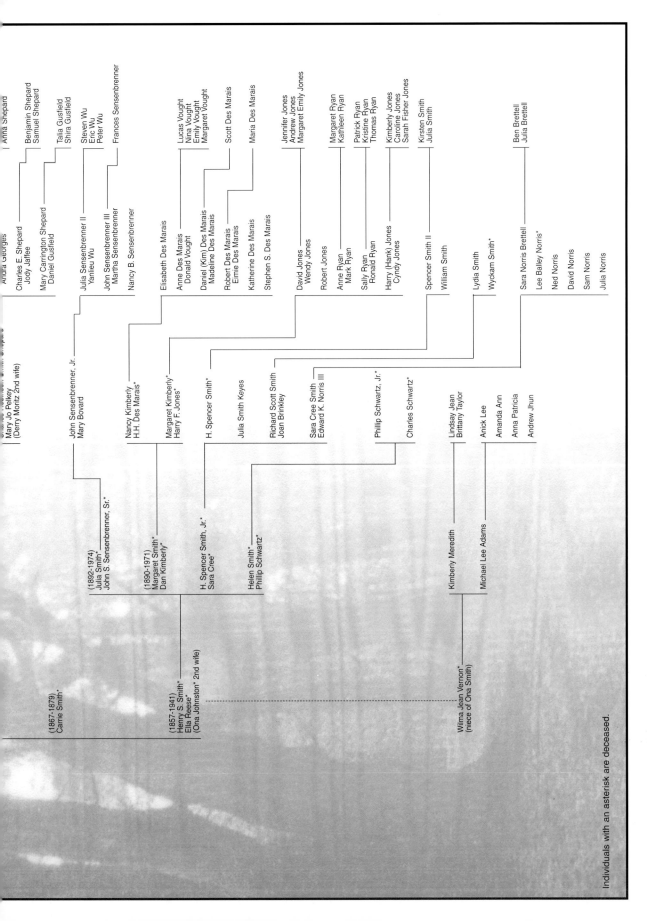

Timeline

1849
The Pail Factory is started in Menasha, Wisconsin, with a $1,000 investment by Nathan Beckworth, Joseph Sanford and C.W. Billings.

1850
The Pail Factory is acquired by Joseph Keyes, Norman Wolcott and Lot Rice. Newlyweds Elisha D. Smith, 23, and Julia Ann Mowry Smith, 21, journey from Woonsocket, Rhode Island, to Menasha, where Elisha goes into the dry-goods business.

1852
Elisha Smith purchases the Pail Factory for $1,200. The one-man shop makes pails, washtubs and other woodenware for local markets.

1857
The Pail Factory survives the Panic of 1857, which bankrupts thousands of businesses across the United States.

1858
Elisha Smith has a spiritual awakening while visiting a church in Chicago. He becomes a devout Christian and gives generously for the next 41 years to the needy and the community at large, becoming the greatest benefactor the City of Menasha has ever known.

1861
The Civil War begins; despite losing the main plant to a fire, the Pail Factory rebuilds and expands, supplying pails and other wooden storage and shipping containers to Union forces during the war.

1862
The Chicago & Northwestern Railroad builds a branch line to Menasha, enabling the Pail Factory to ship its products to Chicago and other distant markets. Elisha Smith closes his dry-goods business to concentrate on the manufacture of woodenware.

1863
Elisha Smith is elected president of the Village of Menasha; reelected the following year, he serves two one-year terms.

1871
The Pail Factory continues to grow, becoming the largest woodenware manufacturer in Wisconsin. It has 250 employees and makes pails, tubs, churns, measures, clothes pins and other wooden products.

1872
With costs rising rapidly in the post-Civil War inflation, and unable to meet its debts, the Pail Factory is thrown into receivership. Elisha Smith's father-in-law, Spencer Mowry, a Rhode Island banker, provides an infusion of cash and reorganizes the business as Menasha Wooden Ware Company. Henry Hewitt, Sr., a local bank executive, becomes president. Elisha Smith becomes general superintendent, regaining the presidency nine years later when Hewitt retires.

1875
Menasha Wooden Ware Company incorporates on May 24 and extends manufacturing operations to central Wisconsin with the rental of a sawmill in Auburndale to make staves for the factory in Menasha.

1878
The pail factory is destroyed by fire.

1885
Phasing out its use of water power, Menasha Wooden Ware purchases one of the nation's first steam-driven electric generators.

1886
In a period of rapid expansion, Menasha Wooden Ware purchases a pail factory at Wrightstown, Wisconsin; a stave mill at Apple Creek, Wisconsin; and 85 acres on Doty Island in Menasha to provide additional drying yards for lumber and staves. The company also opens a sales office in Chicago.

1890
All facilities except the cooperage shop are destroyed by fire and replaced with a brick structure.

1897
Elisha Smith donates 25 acres for the City of Menasha's first public park.

1898
Elisha D. Smith Free Public Library, for which the town's great benefactor donated the site and funds, opens.

1899
Elisha Smith dies at age 72, having built the company into the world's largest manufacturer of turned woodenware. His older son, Charles R. Smith, succeeds him as president.

1900
Menasha Wooden Ware continues to expand, building a stave mill at Ladysmith, Wisconsin.

1903
The company begins acquiring timberlands in the western United States, initially in Idaho.

1904
The company acquires an interest in a large sawmill near Empire, Oregon, one of the first in that state, by purchasing ownership in the Southern Oregon Company.

1916
Charles Smith dies at age 61. Frank D. Lake, a longtime Menasha Wooden Ware employee, succeeds him as president. At the time of Charles's death, the woodenware market is already in decline. Bulk wooden packaging is being supplanted by corrugated boxes, glass jars, metal drums and other newer forms of packaging.

1918
Lake resigns in a disagreement over whether Menasha should diversify into the manufacture of corrugated boxes. Thomas Kearney, a Wisconsin attorney and business consultant, succeeds him as president.

1920
After reorganizing Menasha's operations, Kearney resigns. Willis Miner, a company vice president and friend of the Smith family, succeeds him as president.

1922
To help offset declining demand for woodenware, Menasha forms a subsidiary, Northwestern Wooden Ware Company, Tacoma, Washington, to make Sitka spruce butter tubs, one of the few types of woodenware in demand at the time.

1926
Menasha Wooden Ware is reorganized into two interrelated companies: Menasha Wooden Ware Corporation, which continues in manufacturing and marketing, and Menasha Wooden Ware Company, which buys a portfolio of stocks and bonds to generate income for Menasha shareholders. Willis Miner is president of both.

1927
In a key diversification move, the corporation begins producing corrugated boxes in a converted warehouse.

1929
The corporation begins making "wood flour" at its Tacoma, Washington facility. A fine powder produced from spruce shavings, it is used to manufacture explosives, plastic wood and other products. As demand for woodenware packaging continues to decline, the corporation begins producing pot handles, butcher blocks and other wooden products to keep its workers employed.

1933
Corrugated sales top half a million dollars, while the Woodenware Division is losing money.

1935
Barrel-making is discontinued; the barrel plant is converted to the manufacture of toy and juvenile furniture to keep workers employed. Willis Miner dies. Mowry Smith, Sr., older son of Charles Smith, succeeds him as president of Menasha Wooden Ware Corporation (the operating company). Mowry's brother, Carlton Smith, becomes president of Menasha Wooden Ware Company (the investment company).

1936
Oversized Menasha boxcars, purchased early in the century from a circus, make their last trip, banned because of their wood underframes.

1938
Furniture manufacture is transferred to the stave mill at Ladysmith, Wisconsin.

1939
Expanding "upstream" in the corrugated business, the corporation acquires a 60 percent interest in an Otsego, Michigan, paper mill that manufactures corrugating medium, the squiggly center of corrugated sheet.

1940
The corporation acquires a corrugated box plant in Durham, North Carolina.

1942
When the plant in Ladysmith, Wisconsin, burns, furniture production is transferred to a newly acquired plant in Rockford, Illinois. The corporation produces its last Sitka spruce butter tub.

1945
The corporation acquires a one-fifth interest in John Strange Paper Company of Menasha.

1948
The corporation purchases a one-half interest in a plywood mill under construction at North Bend, Oregon, later acquiring full ownership and running it as Menasha Plywood Corporation.

1949
Menasha Wooden Ware Corporation celebrates its 100th birthday. The Forty-Niner Club (named for the year of the company's founding) begins activities with the recognition of employees with 25 years or more of service; 90 are inducted in this first year.

1952
The corporation discontinues the manufacture of juvenile furniture, converting the Rockford, Illinois, plant to the production of corrugated boxes.

1953
The Smith family forms the Charles R. Smith Foundation, later renamed Menasha Corporation Foundation, to be a focal point for its charitable contributions.

1954
The corporation acquires the timberlands and logging operations of Irwin and Lyons Lumber Co. of Coos Bay, Oregon.

1955
In a time of expansion, the corporation makes a major investment to open a corrugated box plant in Anaheim, California, the box industry's largest plant on the West Coast; acquires full ownership of the Otsego, Michigan, mill; builds a wood flour plant in North Bend, Oregon; and establishes a toehold in plastics, the corporation acquires 51 percent of G. B. Lewis Company of Watertown, Wisconsin.

1956
Al Pierce Lumber Company and its 12,000 acres of Pacific Northwest timberland are acquired, along with Ball Lumber's 1,600 acres.

1957
The corporation ships its last wooden pail, ending its involvement in the woodenware business.

1961
The corporation completes a paper mill in North Bend, Oregon, to make corrugating medium. Mowry Smith, Sr., retires as president of the corporation; he is succeeded by Richard L. Johnson, 45, formerly chief financial officer. Carlton Smith retires as president of the investment company and is succeeded by Mowry Smith, Jr. Fire destroys the Menasha logging camp in Coos Bay, Oregon. Triangle Container Company in Chicago is acquired.

1962
Menasha Wooden Ware Corporation changes its name to Menasha Corporation.

1963
Menasha builds its third wood flour plant in Albany, Oregon. Menasha Corporation opens a corrugated box plant in Medina, Ohio. A new corporate symbol — still in use today — is adopted.

1964
The corporation's box plant in Menasha, in operation since 1927, is destroyed by fire.

1965
The Otsego mill is modernized and expanded with the startup of a new paper-making machine. The company also leases its first computer, an IBM punch-card model.

1966
Menasha Corporation begins operation at a new box plant in Neenah, Wisconsin, replacing the plant in Menasha destroyed two years earlier by fire. In partnership with Green Bay Packaging, Inc., Menasha Corporation acquires Twin Cities Container Corporation of Coloma, Michigan. Menasha opens a Packaging Group Art Center in Menomonee Falls, Wisconsin, to provide customers with graphic design services.

1967
Menasha Plywood Corporation in North Bend, Oregon is closed. The corporate office moves to Neenah.

1968
The corporation acquires a box plant in Tacoma, Washington. Men-Cal Corporation, which gathers and sells paper for recycling on the West Coast, is acquired. The corporation buys Vanant Packaging Corporation (today's Sus-Rap Division).

1969
The company's logging division is closed, along with the Coos River Boom Company responsible for rafting logs in the Coos River. A corrugated sheet plant in Medina, Ohio, is acquired. The Manchester, Iowa, plastic container plant of G. B. Lewis begins operations. The fourth wood flour plant is built in Grants Pass, Oregon. The corporation acquires full ownership of John Strange Paper Company, including its Menasha Paperboard Mill, its Appleton Manufacturing Company subsidiary and its majority interest in Wisconsin Container Corporation (today the Solid Fibre Division of Menasha Corporation).

1970
A newly constructed box plant at Lakeville, Minnesota, begins operations. The first Buckboard expendable pallets are produced. Menasha launches a program to sell its standing timber to local mills in Oregon and Washington.

1971
John Goode becomes the corporation's first "outside" director — that is, neither a member of the Smith family nor a Menasha Corporation employee. Menasha Corporation acquires full ownership of Twin Cities Container Corporation. The corporation acquires a 33 percent interest in Poly Hi Inc. of Fort Wayne, Indiana, which processes and markets ultra-high molecular weight (UHMW) polyethylene; full ownership is acquired over the next six years.

1972
The corporation acquires a corrugated box plant in Hartford, Wisconsin, and forms the Color Division at Hartford to produce corrugated boxes with high-quality color printing.

1973
The Monticello, Iowa, plastic container plant of G. B. Lewis Company begins operations. Menasha corporation opens a Convoy pallet plant in Menasha, the only plant in the industry devoted exclusively to making plastic pallets.

1974
The corporation acquires Crown Corrugated Containers (a predecessor of today's Yukon Packaging Division) of Greensburg, Pennsylvania. The fifth wood flour plant is built in Centralia, Washington.

1975
Full ownership of G. B. Lewis Company is acquired; Lewis is divided into two Menasha divisions — LEWISystems and Molded Products.

1977
The Ohio Valley Container Corporation is acquired and becomes Menasha Corporation's Mt. Pleasant, Tennessee, plant. Carlin Container Corporation of Roselle, New Jersey, is acquired.

1980
Dare Pafco Products Company of Urbana, Ohio, is acquired as a LEWISystems plant. The acquisition of International Paper Company's Yukon box plant at Youngwood, Pennsylvania, provides expansion for the nearby Greensburg plant.

1981
Menasha Corporation sells the North Bend paper mill, Anaheim box plant and Portland and Eugene, Oregon, secondary fiber facilities to Weyerhaeuser Company. The operating corporation and investment company merge, ending a separate portfolio of stocks and bonds. Richard Johnson retires as CEO; he is succeeded by Donald C. "Tad" Shepard, Jr., 58, a great-grandson of Elisha Smith who has been with Menasha Corporation since 1950. Scranton Plastics Laminating Corporation of Scranton, Pennsylvania, is acquired.

1982
The company acquires Vinland Corporation (web-printed paper and plastic film products) of Neenah, Wisconsin, which later merges with the Neenah Printing Division. Menasha Corporation's truck fleet is converted to an interstate common carrier, Menasha Transport, Inc. The Land & Timber Division ships its first timber to Asia, launching a highly successful export program.

1984
Traex Corporation (reusable plastic products for the food service industry) of Dane, Wisconsin is acquired. A corrugated box is plant built at Olive Branch, Mississippi.

1985
Mid America Tag & Label Co. (product identification and merchandising tags and labels) of Neenah, Wisconsin, is acquired. Poly Hi forms a venture with Tsutsunaka Plastic Industry Co. Ltd. of Japan, Menasha Corporation's first international joint venture.

1986
Menasha Corporation acquires Neenah Printing Company and its subsidiary, Oshkosh Printers, Inc. Murfin, Inc. (web-screen printer of label and identity products) of Columbus, Ohio, is acquired.

1987
The acquisition of Dripcut Corporation's product line expands Traex Division's product offerings for restaurants. Menasha Corporation acquires West Tape & Label Co. of Denver, Colorado. Waldron, Inc., is acquired and becomes the Marysville, Washington, plant of the Wood Fibre Division. Mid America Industrial Division is created to produce tags and labels for industrial markets. Menasha enters a joint venture with Caffell Brothers of Portland, Oregon, to build a sawmill in Coos Bay, Oregon.

1988
FL Industries' Thermotech Division (precision injection molding of thermoplastics and engineered resins) of Hopkins, Minnesota, is acquired.

1989
Colonial Container Company (corrugated boxes) of Green Lake, Wisconsin, is acquired. Tad Shepard retires as president and CEO; he is succeeded by Walter H. Drew, 54, recruited from Kimberly-Clark Corporation. The Mid America Division expands through the acquisition of Labelcraft Corporation of Farmingdale, New Jersey. Menasha sells the sawmill joint venture Coos Bay Cedar to Weyerhaeuser.

1990
Denney-Reyburn Co. (tags and labels) of West Chester, Pennsylvania, is acquired. North Star Container, Inc., is acquired and becomes the Brooklyn Park, Minnesota, box plant.

1991
The Printed Systems Division is formed, combining the Mid America Industrial Division and the Denney-Reyburn plant in Tempe, Arizona; the Denney-Reyburn plant in West Chester, Pennsylvania, is closed.

1992
Menasha P.O.S. (point of sale) is formed to sell promotional items manufactured by the Color Division, Neenah Printing and the Mid America Division. Walter Drew resigns; Tad Shepard resumes the presidency on an interim basis.

1993
Robert D. Bero, 52, vice president-plastics, is elected president and CEO, succeeding Shepard. The corporation purchases the U.S. operations of Solidur Deutschland GmbH, the leading European producer of UHMW sheet, and merges those operations with domestic operations of Poly Hi to form the Poly Hi Solidur Group. DuraPAK, a total returnable packaging systems business, is established with production in Danville, Virginia, and technical and sales support in Cincinnati, Ohio. Menasha Corporation and Libla Industries, Inc., form a joint venture, First National Pallet Rental, Inc. in St. Louis, Missouri, to rent pallets to the food and grocery industry. New Jersey Packaging (pressure-sensitive and heat-seal labels for the pharmaceutical and health care industries) of Fairfield, New Jersey, is acquired.

1994
The Color Division changes its name to DisplayOne, reflecting its growth in the design and manufacture of point-of-purchase displays for consumer product companies. LEWISystems expands with the construction of a plant in Urbana, Ohio.

1995
Donray Company (foam-cushioned packaging, including corrugated cartons with cushion inserts) of Mentor, Ohio, is acquired. Insert Corporation de P. R., or Insertco, (inserts and outserts for pharmaceutical companies) of San German, Puerto Rico, is acquired. In separate transactions, two corrugated box businesses — Mid South Packaging of Cullman, Alabama, and Southwest Container Corporation of Phoenix, Arizona — are acquired.

1996
WolPac, Inc., a Madison Heights, Michigan, company that designs, engineers and prototypes material handling systems used by the automotive industry, is acquired. Menasha Corporation creates ORBIS returnable packaging business by combining the LEWISystems, Convoy Plastic Pallets, DuraPAK and Donray Divisions. Menasha Corporation forms the Promo Edge Division, combining the Mid America Tag & Label, DisplayOne and point-of-sale business; the new division provides a full range of in-store promotional products and related services to consumer product companies. A corrugated sheet plant opens in Erie, Pennsylvania. Middlefield Container Corporation of Middlefield, Ohio, is acquired.

1997
Menasha Corporation purchases the operations of Solidur Deutschland GmbH outside the United States, merging them into Poly Hi Solidur Group. Menasha Services business is established. The Molded Products Division is sold.

1998
Montec Plastics of Monson, Massachusetts, formerly a Thermotech plant, is established as a separate division. The Land & Timber Division plants its 30 millionth seedling on company lands in Oregon and Washington.

1999
Menasha Corporation celebrates its 150th birthday.

Acknowledgments

We gratefully acknowledge the contributions of the following individuals and organizations in sharing their recollections or providing photographs or other materials for this book:

Frank Albert
John Alexander
Dawn Arent
Dow Beckham
Robert Bero
Norman Brown
Bruce Buchanan
Jeff Burleigh
Mark Cane
Richard Clarke
Robert Conrad
Lyle Crandall
Peter DeRossi
Anne Des Marais
Nancy Des Marais
Walter Drew
Kirby Dyess
Ursula Fairbairn
Boyd Flater
Laurie Franczak
Milton Frater
Evan Galbraith
Katharine "Kig" Gansner
Susan Gosin
William Griffith
Ric Hartman
Wendy Heenan
Virginia Johnson
Harry Jones III
B. H. "Keg" Kellogg
Kenneth Kiesau
John Kline
Barb Koenig
Carl Kraus

Steven Kromholz
Tom Kukuk
Langston Corp.
William Lansing
Lynn Larson
Mowry Mann
Bernard "Mac" McCarragher
Dorothy McQuillan
Menasha Historical Society
Menasha Public Library
   (Elisha D. Smith Public Library)
Lucile Miller
Helen Morrell
Dale Mosier
Munroe Studios
Mike Muntner
Jeff Murphy
Dan Murton
Neenah Historical Society
Neenah Public Library
Jeff Nyman
Sara Otto
Michelle Papierniak
Ron Patton
Jill Paull
Kristine Pavletich
Raeburn "Rae" Peppler
Thomas Prosser
Rob Riley
Donald Riviere
Jack Rohde
Jerry Roovers
Lee Rottinghaus
Dave Rust
James Sarosiek

H. E. "Rusty" Sattler
Susan Schaefer
Allan Schenck
John Schmerein
Bruce Schnitzer
Walter Sellnow
John Sensenbrenner, Jr.
Charles R. Shepard
Donald C. "Tad" Shepard, Jr.
Donald C. "Buzz" Shepard III
Jane Shepard
Sylvia Shepard
Timothy Shepard
William Shepard
Clark R. Smith
Curtis Smith
Mowry Smith, Jr.
Mowry "Trip" Smith III
Oliver Smith
John Snyder
Allen Stinchfield
Kevin Stengl
Walter Stommel
Henry Suess
Susan Surendonk
Kathy Swick
James and Monica Taylor
Donald Turner, Jr.
Lucas Vought
Michael Ward
Greg Wenkman
Thomas Williscroft
Bill Zimmer

INDEX

Bold listings indicate illustrated material.

A

Albert, Frank, 65-66, 165
Alexander Art Corporation, 135
Alexander, John, 165
Allegan County News & Gazette, 69
American Automobile Association, 131
American Board of Commissioners for Foreign Missions, 28
American Can Company, 92
American Cotton Oil Corporation, 151
American Federation of Labor, 61
American Forest & Paper Association, 61
Amtrak, 135
Anaheim, California, corrugated container plant. *See under* Menasha Corporation
Anadromous, Inc., 101
Anderson, Delores, 97
Anderson, Dorothy, **98**
Anheuser-Busch Companies, Inc., 141 **141**
Appleton Manufacturing Company, 71, 101
Arent, Dawn, 165
Arkitoy, 125
Associated Brewing Company, 102

B

Ball Lumber Company, 79
Bardeen, George, 69
Barnard College, 108
Beaton, Otto, **42**
Beaudo, Joe, **22**
Beckham, Dow, 79, 84, 165
Beckworth, Nathan, 22
Beloit College, 29, 43
Beloit Corporation, 89
Bender, Phyllis, **98**
Berlin, 20
Bero, Robert D., 11, **11**, 15, 92, 121, **127**, 129, 133, 135, 137, 139, 142, 145, **151**, 152-157, **154**, **156**, **159**, 165
Billings, C. W., 22
Bingham newspaper chain, 9
Black & Decker Manufacturing Company, 123
Boyd, William, 32, 34
Briggs & Stratton Corporation, 142

Briggs, Robert M., 73, **108**, 109
Britzke, Dorothy, **98**
Brown, Norman, 138, 165
Buchanan, Bruce, 165
Buchta, Don, 142
Burger King, 130
Burleigh, Jeff, 165
Burlington Northern Railroad, 135
Bush, George, **111**

C

California, University of at Berkeley, 99
Cane, Mark, 135, 165
Celanese Corporation, 101
Chicago & Northwestern Railroad (the Northwestern), 23-24, 40
Clarke, Richard, **11**, 100-101, 152, 165
Coca-Cola Company, 142
Colonial Container Company, 114
Columbia Records, 72
Conrad, Robert, 151, 165
Container Corporation of America, 59, 69, 71, 140
Convoy Plastic Pallets, 101, **125**, 133
Coos River Boom Company, 82-84
Crandall, Lyle, 165
Crown Corrugated Containers, 101

D

Dare Pafco Products Company, 101
Dart Group, 9
Deere & Company, 126
DeRossi, Peter, 165
Des Marais, Anne (great-great-granddaughter of Elisha and Julia Smith), 13, **157**, 165
Des Marais, Nancy (great-granddaughter of Elisha and Julia Smith), 10, 40, 43, 165
Diamond Crystal Salt Company, 102
Distilling & Cattle Feeding Company, 151
Doane, Dr., 17, 20
Dombeck, John, **98**
Donray, 133
Doty Island, 19, 23, 34
Dow Jones Industrial Average, 151
Downer College, 34
Drew, Marion, 120
Drew, Walter H., 107, 110, 119-121, **120**, 165
Dripcut, 129-130
Dun & Bradstreet, 20
Dunsirn, Bob, 142
DuPont Company, 135
DuraPAK, 125, 133
Durham Container Company, 70
Dyess, Kirby A., **11**, 12, 101, 165

E

Elisha D. Smith Free Public Library, 10, **16**, 17, **34**, 35, 165

F

Faegre & Benson, 10
Fairbairn, Ursula, 165
Fibreboard Corporation, 102
Fibre Box Association, 108
Fillmore, Millard, 17
Firgens, Elder "Al," 105
First Congregational Church of Menasha, 21, **27**, 43
Flater, Boyd, 125, 165
FL Industries, 131
Forest Products Group. *See under* Menasha Corporation
Fort Wayne Corrugated Paper Company, 52, 53
Fox Cities Children's Museum, **158**
Fox, Ed, 61
Fox River, 19, 23, 30
Franczak, Laurie, 165
Frater, Gordon, 125
Frater, Milton, 125, 165

G

Galbraith, Evan, **11**, 118, 121, 165
Gamble, Ray, 58-59
Gansner, Katharine G. "Kig" (great-great-granddaughter of Elisha and Julia Smith), 10, **11**, 12, 13, 64, 71, 165
Gansner, Nick (great-great-great-grandson of Elisha and Julia Smith), **13**
General Electric Company, 125, 151
General Mills, Inc., 72, 96, 139
General Motors Corporation, 61, 126
Gillette Company, 139
Globe Bank, 18
Goeser, Clarence, 97
Golden, Ray, **98**
Golden Valley Groves, 101
Goldman Sachs & Co., 102
Goode, John, 100
Gosin, Susan (great-great-granddaughter of Elisha and Julia Smith), **159**, 165
Greene, David, 69-70
Griffith, William, 91-92, 93, 107, 165
Grinnell College, 29

H

Habitat for Humanity, 158
Hardee's, 130
Harley-Davidson Inc., 126
Harney, Richard, 23
Hartford, Wisconsin, plant. *See under* Menasha Corporation
Hartman, Ric, 165

Hawkins, John, 78-79
Heenan, Wendy, 165
Helgerson, Karen, **159**
Hewitt, Henry, Sr., 29, 33
Hinton, George, **52**, 60-61, 93
Hintz, Dick, 82
Hoffman, Dustin, 123
Hotchkiss School, 108

I

Intel Corporation, 12, 101
Internal Revenue Service, 92
International Business Machines Corporation (IBM), 105
International Paper Company, 59, 69, 140
Irwin and Lyons Lumber Company. *See under* Menasha Corporation
ITT Corporation, 131

J

Jack, Neal, **98**
Jankowski, Joe, **98**
Johnson, Gregg, 93
Johnson, Richard L., 11, 63, 87, 91-104, **91**, **92**, **95**, **98**, **100**, 109-110, 112, 114, 118, 119, 120
Johnson (S. C.) Company
Johnson, Timothy, 93
Johnson, Virginia, 92, 93, 165
Jones, Harry F. III (great-great-grandson of Elisha and Julia Smith), 10, **11**, 165

K

Kass, Alvin, 67
Kass, Mary, 67
Kearney, Thomas M., 49
Kellet, Marie, 97
Kellogg, Breadon H. "Keg," 52, 97, **98**, 165
Kelly, Ed, **63**
Keyes, Joseph, 23
Kiesau, Kenneth, 165
Kiesling, Del, 127
Kimberly-Clark Corporation, 67, 72, 89, 96, 110, 119-120
Kimberly, D. L., 73
Kline, John, 165
Koenig, Barb, 165
Kraft Foods, Inc., 139, 145
Kraus, Carl, 118, 165
Kromholz, Steven S., 159, **159**, 165
Kukuk, Tom, 165

L

Labelcraft Corporation, 142
Lake, Frank D., 40, 48-49, 59

Lakeville Packaging Division. *See under* Menasha Corporation
Lake Winnebago, 19, 21, 28
Land O'Lakes Creamery Association, 56
Langston (Samuel M.) Co., 59, 165
Lansing, William, 77, 79, **93**, 94, 99, 112, 118, 165
Larson, Lynn, 165
Lauer, John, **11**
Lawrence University, 29, 43
Lewis, G. B., 124
Lewis (G. B.) Company, 89, 94, 123-126, **124**. *See also* Menasha Corporation, LEWISystems
Lewis, R. E., 124
Life, 84
Life Savers, 50
Lismire, Nancy, **98**

M
MacKenzie, Len, **98**
Manders, Ernest C., 87, 93
Mann, Mowry, 165
Marathon Corporation, 11, 91, 92, 93
Marsh & McLennan, Inc., 12
Massey, Clem, 45
Massey, Elmer, **45**
Material Handling Group. *See under* Menasha Corporation
Mayer (Oscar) & Co., 72, 96, 142
Mayflower, 17
McCarragher, Bernard "Mac," 91, 93, 97, **98**, 104, 110, 114, 121, 165
McDonald's Corporation, 130
McQuillan, Dorothy, 165
Mechler, William, **24-25**
Menasha 1980 Corporation, 104
Menasha Art Center. *See under* Menasha Corporation
Menasha, City of, 9, **16**, 17, **19**,
Menasha Corporation (*see also* Menasha Wooden Ware Company; Menasha Wooden Ware Corporation): advertising, **57**, **67**; Anaheim, California, corrugated container plant, **60**, 60-61, 71-72, 84, 88, 89, 94, **96**, 104, 112, 152; board of directors, **11**, 11-12, 73, 100-101, **100**; boxcars owned by, **24-25**, **41**; butter tub business, 55-58, **56**, **57**, **58**, 78; charitable contributions of, 12; children's history book of company, **12**, 13; Coloma Packaging Division, 96, 159; company origins, 9, 22; Convoy pallet business, 101; corrugated container business, 10, 59-61, 65-73, 94; DisplayOne, 141, 142; dividend payments, 15, 114, 152; Erie Packaging, 145; Farmingdale, New Jersey, plant, 142; fires at company facilities, 26, 74, **95**, 97-98; Forest Products Group, 26, 41-43, 77-89, 99-100, 118, 137, 156; growth strategy, 15, 101-104, 129, 152-157; Hartford, Wisconsin, plant, 101, 140-141; insolvency of in 1870s, 29; *Log*, 59; Irwin and Lyons Lumber Co., 78-79, **81**, 82, 84; Lakeville Packaging Division, 101; LEWISystems, 89, 94, 123-126, 133; longevity of, 10; manufacturing facilities on Fox River, **16**, 32, 41, 50-51; Material Handling Group, 123, 133, 135; Medina, Ohio, plant, 96; Menasha Art Center, **137**, **152**; Menasha Services Division, 133, **155**; Menasha Transport, Inc., 117-118, **117**; Men-Cal subsidiary, 101; Middlefield Packaging, 145; Mid South Container Division (Cullman, Alabama), 145; Mission Statement, 113, 152; Molded Products Division, 126; Montec Plastics Division, 131, **132**, 133; Murfin Division, 114, **130-131**, 131; Neenah Packaging Division, 98-99, 108-109; Neenah Printing Division, **14**, **15**, 114, **137**, 138; New Jersey Packaging, **145**; North Bend, Oregon, paper mill, 72, 73, 78, 84-89, **87**, **88**, 89, 93, 94, 104, 112; Olive Branch Division, 141; ORBIS, 101, **122**, 125, 133-135, **134**, 145, 154-156, **154**; Packaging Group, 10, 59-61, 65-73, 94, 145-149, **146-147**; Paperboard Division (Otsego, Michigan, paper mill), **68-69**, 69-70, 72-73, 89, 96-97, 100, **116**, 117, 123, 137-138; plastics business, 11, 89, 123, 129, 135; plywood mill, 78, **78**, 79, 94, 95, 109; Poly Hi Solidur, **123**, **126**, 126-129, 156, **156**; Polymer Technologies Group, 123, 129-133; Promo Edge, 139-142, **139**, **140**, 145, 156; Promotional & Information Graphics Group, **138**, 138-145; St. Cloud, Minnesota, plant, 145; Solid Fibre Division, 66, 71, **144**, 147-148, 149; Sus-Rap Division, 96, 147, 148-149, **149**; Thermotech Division, 114-117, **115**, 131-132; toy and juvenile furniture business, 58, 74; Traex Division, 114, **128**, 129-131; woodenware business, 22-23, 26, 31-32, 49-50, 74; wood flour business, 58-59, 109; Yukon Packaging Division, 101
Menasha Corporation Foundation, 35, 158-159
Menasha Historical Society, 45, 165
Menasha Transport, Inc., 117-118, **117**
Menasha Wooden Ware Company, 29, 30, 35, 39, 53, 54-55, 63; butter tub business, 55-58; candy pail business, 32, 50
Menasha Wooden Ware Corporation, 53, 54-55, 63
Men-Cal Corporation, 101
Mid America Tag & Label Co., 114, 142
Miller, Lucile, 93, **98**, 165
Miner, Reverend H. A., 21, 49
Miner, Willis, 21, 40, 48, **48**, **49**, 53, 59, 63
Minnesota Mining & Manufacturing Company, 108, 141
Molded Products Division. *See under* Menasha Corporation
Montec Plastics Division. *See under* Menasha Corporation
Moody, Dwight L., 28
Moore Mill and Lumber Company, 81
Morgan Guaranty Trust Company, 100
Morgan Stanley & Co., 102
Morrell, Helen, 165
Mosier, Dale, 129, 165
Mowry, Spencer (father-in-law of Elisha Smith), 17, 18, **18**, 23, 29, 30, 32, **33**, 37
Munroe Studios, 165
Muntner, Mike, 52, 67, 97, 99, 165
Murfin, Clif, 131
Murfin, Inc. *See* Menasha Corporation, Murfin Division
Murphy, Jeff, 165
Murton, Dan, 165

N
National Bank of Menasha
National Biscuit Company, 50
National Industrial Recovery Act of 1933, 61
Neenah, City of, 19
Neenah Citizen and Menasha Register, 27, 34
Neenah Historical Society, 165
Neenah Printing. *See under* Menasha Corporation
Neenah Public Library, 165
New Dramatists, 159
New York Stock Exchange, 103
Nichimen Corporation, **154**
Nighswonger, Al, **86**
North Bend, Oregon, paper mill. *See under* Menasha Corporation
Northwestern Railroad. *See* Chicago & Northwestern Railroad
Nyman, Jeff, 165

O
Oglebay Norton Company, 102
Olinkraft Inc., 103
ORBIS. S*ee under* Menasha Corporation
Oregon Public Utility Commission, 82, 84
Oshkosh Printers, Inc., 114, 138
Otsego, Michigan, paper mill. *See* Menasha Corporation, Paperboard Division
Otto, Sara, 165

P
PACCAR, Inc., 135
Pacific Power and Light Company, 87
Packaging Group. *See under* Menasha Corporation
Packard Bell, Inc., 72
Page, Nelson, 93, **95**
Pail Factory (predecessor of Menasha Corporation), **19**, 22, 24, 152
Panic of 1857, 23
Papierniak, Michelle, 165
Patton, Ron, 165
Paull, Jill, 165
Pavletich, Kristine, 165
Pennekamp, Guenther, 127-128, **127**
Peppler, Raeburn "Rae," **98**, 165
PepsiCo, 142
Peytona, 18
Pierce (Al) Lumber Company, 79
Pillsbury Company, 108, 139
Pittsburgh Paint, 70
Point of Purchase Advertising Institute, 141
Pre-cut Sawmill, **154**
Princeton University, 38, 43, 47, 53, 54, 73, 108
Promo Edge. *See under* Menasha Corporation
Promotional & Information Graphics Group. *See under* Menasha Corporation
Prosser, Thomas, **11**, **159**, 165

R
Radford, Peter (great-great-grandson of Elisha and Julia Smith), **159**
Reed, Curtis, 18
Rice, Lot, 23
Riley, Rob, 165
Ripon College, **28**, 29

Riviere, Donald, 110, **112**, 113, 114, 120, 129, 133-135, 165
Rohde, Jack, 165
Roovers, Jerry, 131-132, 165
Rottinghaus, Lee, 165
Russell, Tom "Tucker," 51
Rust, Dave, 135, 165

S
Sanford, Joseph, 22
Sarosiek, James, **159**, 165
Sattler, H. E. "Rusty," 71, 73, 165
Schaefer, Susan, 165
Schenck, Allan, **102**, 117, 165
Schmerein, John D., 40, 165
Schnitzer, Bruce W., **11**, 12, 91, 100, 104, 165
Schultz, Audrey, **98**
Scranton Plastics Laminating Corporation, 114, 127
Sears, Roebuck & Co., 70, 126
Sellnow, Walter, 97, 148, 165
Sensenbrenner, John, Jr. (great-grandson of Elisha and Julia Smith), 108, **157**, **159**, 165
Sensenbrenner, Mary, **157**
Shepard, Charles R. (great-grandson of Elisha and Julia Smith), 107, 110-111, 165
Shepard, Chester, **55**
Shepard, Donald C., Sr., **38**, **52**, 53, **54**, 55-56, **55**, 57, 61, 63, 73, **100**, 108
Shepard, Donald C. "Tad," Jr., (great-grandson of Elisha and Julia Smith), 12, 44, 71, 73, 74, 77, 84, **92**, 93, 94, **95**, **98**, **100**, 104, **111**, 127, 138, 141, 148, 151, **159**, 165; as president of Menasha Corporation, 108-120; retirement, 120-121
Shepard, Donald C. "Buzz" III (great-great-grandson of Elisha and Julia Smith), 10, **11**, **121**, 165
Shepard, Jane Steinborg, 108, **121**, 165
Shepard, Lydia (great-great-great-granddaughter of Elisha and Julia Smith), **158**
Shepard, Peter (great-great-great-grandson of Elisha and Julia Smith), **13**
Shepard, S. Sylvia (great-great-granddaughter of Elisha and Julia Smith), **121**, **159**, 165
Shepard, Sylvia Smith (granddaughter of Elisha and Julia Smith), **38**, 43, 53, 57, 73, 107
Shepard, Timothy C. (great-great-grandson of Elisha and Julia Smith) **11**, 12, **121**, 165

Shepard, William A. (great-great-grandson of Elisha and Julia Smith), 10, **11**, 11, **121**, 135, 145, **148**, 165
Smith, Barney & Co., 102
Smith, Carlton R. (grandson of Elisha and Julia Smith), 43, **47**, 48, **48**, **52**, **53**, 53-54, **54**, 57, 61, 73, **76**, 91, 93, 107
Smith, Carrie (daughter of Elisha and Julia Smith), 24
Smith, Charles R. (son of Elisha and Julia Smith), 18, 24, 30, 34, 35, 37-44, **37**, **38**, 49, 50, 77, 81, 107, 113, 158; broom-handle factory owned by, 38-39, 40, **43**; death of, 44; failure to groom a successor, 47-48
Smith, Clark R. (great-grandson of Elisha and Julia Smith), **11**, **106**, **159**, 165
Smith College, 43
Smith, Curtis N. (great-great-grandson of Elisha and Julia Smith), 11, **11**, 13, 165
Smith, Elisha D. (founder of Menasha Corporation), 9, 10, 11, 12, 13, **17**, 18, **21**, 22, 26, 30, 49, 73, 104, 107, 124, 151, 152, 157; death of, 35; dry-goods store owned by, 20, 23, 24; generosity of, 20, 27-29, 34-35, 158; marriage to Julia Ann Mowry, 17; relationship with father-in-law, Spencer Mowry, 18, 29, 33; religious faith of, 27-29
Smith, Ella Reese (daughter-in-law of Elisha and Julia Smith), 43
Smith, Henry (son of Elisha and Julia Smith), 24, 38, 39-40, 44, 48; death of, 44
Smith, Isabel Rogers (daughter-in-law of Elisha and Julia Smith),
Smith, Jan, **98**
Smith, Jane (daughter of Elisha and Julia Smith), 24, 40, 73; death of, 44
Smith, Jennie Mathewson (daughter-in-law of Elisha and Julia Smith), 43
Smith, Julia Ann Mowry (wife of Elisha Smith), 17, 18, 19, **20**, **21**, 22, 23, 24, 29; death, 35
Smith, Lawton (great-grandson of Elisha and Julia Smith), 109
Smith, Mary (daughter of Elisha and Julia Smith), 24
Smith, Mowry, Sr. (grandson of Elisha and Julia Smith), 43, 44, **47**, 48, **48**, **52**, **53**, 53, 54, 57, 59, 61, **62**, **63**, 63, 69-70, **70**, 71, 73-74, 77, 79, 91, 92, **92**, 93, 107, 109, 113, 125, 147, 158; accomplishments of, 89; death of, 89, 97; love of people, 53, 64-65

Smith, Mowry, Jr. (great-grandson of Elisha and Julia Smith), **8**, 9, 10, 37, 43, 50, 58, 63, 73, 79, **92**, 93, **98**, 100, **100**, **106**, 109, **109**, 125, 157, **157**, 165
Smith, Mowry "Trip" III (great-great-grandson of Elisha and Julia Smith), **8**, 9, 13, 15, 165
Smith, Oliver C. (great-grandson of Elisha and Julia Smith), **8**, 9, **11**, 93, 98, **100**, 109, **159**, 165
Smith, Onnie (great-great-granddaughter of Elisha and Julia Smith), **159**
Smith Park, 9, 34, **34**; Smith family picnic held in park, 12, 34, **35**, **150**
Smith, Pierce (great-great-grandson of Elisha and Julia Smith), **8**, 9
Smith, Sylvia (granddaughter of Elisha and Julia Smith), 43-44, 48
Smith, Tamblin C. (great-grandson of Elisha and Julia Smith), 73, 93, **100**, 101, 109
Smith, Theda Peters, **54**
Snyder, John, 133-135, 165
Solidur Deutschland GmbH, 127-129
Splash dam logging, 82-84, **83**
Standard & Poor's 500 Index, 15
Stanford University, 109
Stengl, Kevin, 165
Stinchfield, Allen P., 79-81, 93, 94, 165
Stinchfield, Ellen, 79-81, 87, 99
Stommel, Walter, 51, 52, 64-65, 74 **75**, **157**, 165
Strange (John) Paper Company, 60, 71, 89, 101, 148, 158
Strategic Decisions Group, 152
Suess, Henry, 165
Suess, Ralph, 61, 93
Suess, Roman, **100**
Sunkist Growers Cooperative, 72
Surendonk, Susan, 165
Swick, Kathy, 165

T
Taylor, James, 165
Taylor, Monica, 165
Tenryu Lumber Ltd., **154**
Thermotech. *See under* Menasha Corporation
Thomsen, David, 138
Traex Corporation. *See* Menasha Corporation, Traex Division
Triangle Container Company, 96, 97
Tsutsunaka Plastic Industry Co. Ltd., 127
Turner, Donald, Sr., 47, **52**, 53, **54**, **55**, 55-56, 59, 61, 63, 73, 77, 79, 85, 91, **100**

Turner, Donald, Jr., 56, 165
Tuskegee University, 29

U
Ultra-high molecular weight (UHMW) polyethylene, 123, **126**, 126-127
Uneeda Biscuits, 50
Union Carbide Corporation, 72
United Parcel Service, **144**, 148
United States Leather Company, 151
U.S. Paper Mills Co., 71

V
Vanant Packaging Corporation, 96, 149
Van Antwerpen, Martin, 149
Vinland Corporation, 114
Vought, Emily (great-great-great-granddaughter of Elisha and Julia Smith), **10**, **157**
Vought, Lucas (great-great-great-grandson of Elisha and Julia Smith), 165

W
Waite, Julia Shepard (great-great-granddaughter of Elisha and Julia Smith), **121**
Wal-Mart Stores, Inc., 145
Ward, Michael, 165
Washington, University of, 99
Wenkman, Greg, 129, 165
Weybright, Red, **85**
Weyerhaeuser Company, 73, 78, 87, 89, 104, 107, 112
Widmann, Richard, 93
Williscroft, Thomas, 89, 165
Wilpolt, Henry, **24-25**
Wilson, Woodrow, 38
Wilton College, 29
Winnebago Indians, 19
Wisconsin Container Corporation, 66, 71, 97, 101, 148
Wisconsin Northern Railroad, 40-41
Wisconsin, University of, 49, 125
Wolcott, Norman, 23
Woodenware Manufacturers Association, 32
Worden, Charles, 26

Y
Yale University, 53, 108, 109, 111
Young, John, **52**
Yukon Packaging Division. *See under* Menasha Corporation

Z
Zenefski, Pam, **98**
Zimmer, Bill, 165